KNOWING THE UNKNOWN - III

Challenges of Technology
- Past, Present and Future

Manohar Lal, Ph.D.

Library of Congress Catalog Number: 2001012345
ISBN: 978-0-9826809-2-6

MRLT, LLC, Tulsa
Printed in the United States of America

Table of Contents

ACKNOWLEDGMENTS

Challenges of Technology is the third book in the series entitled Knowing the Unknown. It addresses the question, "What am I doing?" The first book in the series, Mysteries of Life, addresses the question, "Who am I?" The second book in the series, Mysteries of The Universe, addresses the question, "Where am I?"

I want to acknowledge gratefully the help I received from all the sources, which are too numerous to list. I have listed the names of many distinguished persons in the book, who have contributed directly or indirectly in some manner to the development of technology. With the popularity of the Internet, one can easily find further information about their work by typing a few words on a search engine.

I also want to acknowledge help from my daughter-in-law, Ritu, who designed the cover page. It would not have been possible to write this book if my friends, my children, grandchildren, and my wife Rani had not inspired me to do so. I would like to dedicate this book to all of them. I dedicate this book to my wife, Rani, who has been my main inspiration in life.

Preface

In this book, we take the reader on an exciting journey that unfolds the mysteries of technology. During this journey, we visit and cross various milestones in the history of technology that have changed our lives forever. The journey through technology reveals what we have accomplished through the application of science. We examine the current frontiers and the future of technology. We list and discuss the top ten challenges in the field of technology.

As we become familiar with the ways, in which technology affects our lives and allows us to control our environment, we ask the following questions: Can we develop new resources and new materials that possess the properties of our choice? Can we develop new means of communication and design intelligent computers, intelligent control systems, and robots? We address these questions in this book.

We observe that developments in technology parallel progress in science. We see how technology has helped us harness natural resources, control our environment, communicate with each other, and extend our decision-making abilities through computers. We also discover that despite tremendous advances in technology, several challenges remain.

This is not just another book on technology. It integrates various technological developments derived from different scientific concepts. It explains various technologies in simple terms. It also takes an objective look at how technology has changed our lives. This book will heighten readers' interest in technology. It will also motivate and encourage young scientists and engineers to seek new technology. Finally, it can serve as a textbook for students of the history of technology.

My credentials for writing this book include my diverse educational and professional backgrounds. I received my MS and PhD degrees in electrical engineering in 1960 and 1963 from the University of Illinois, Champaign-Urbana, Illinois. I have taught and/or conducted research in science and technology for over forty years at several universities and at an industrial research center. This experience provided opportunities for research in diverse areas of science and technology.

Bon Voyage!
Manohar Lal

Chapter 1
Technology – Gift of Basic Science

1.1 Introduction

We now embark on a journey through technology. My second book in the series entitled ***Knowing the Unknown***, *Mysteries of the Universe*, discusses various discoveries in science. We start our present journey by revisiting each milestone of science that led to the development of technology. We explore how technology evolved from the earlier discoveries in basic science.

We start with the development of machines from the laws of mechanics discovered by Newton, heat engines from the laws of thermodynamics, and wireless communications from the laws governing electric and magnetic phenomena unified by Maxwell. In the next chapter, we shall continue with technology based on modern science and physics.

1.2 Technology & Science

The impact of technology on society and the environment goes back many millennia. A theory concerning Homo sapiens states that control over fire allowed early humans to cook meat, thereby leading them to develop lighter jawbones. As a result, they had to work less on chewing, which eventually led to the development of bigger brains!

In terms of the human journey, next came the techniques of agriculture, domestication of animals, buildings, metallurgy, transport, crafts, and writing. All of these advances had an enormous impact on societies and civilizations. With scientific knowledge, we developed modern technologies and the means of systematically exploiting the gifts of nature and science over the last four centuries. These technologies are also much more energy-intensive than the earlier technologies

The importance of technology in our lives is self-evident. We now live in a world in which we can contact anybody, anywhere, at any time. We can watch any event, anywhere in the world, at any time we choose. We can remotely control and manage machines that perform our hazardous chores for us.

Technology has enabled us to live in such a world. It has touched the life of every individual on the planet. Most of us now take such applications for granted. The younger generation cannot even imagine life without machines, electricity, transportation, communication devices, computers, and other modern technology.

Technology owes its existence to science, since it conceives what science perceives. Science formulates theories explaining phenomena concerning energy fields, matter, space, information, and time. Technology puts these theories into practice for the benefit of humanity. Technology thus finds applications for the scientific concepts, which makes our lives more comfortable.

We observe how applications of various scientific principles lead to the development of new technologies. Newtonian physics and thermodynamics led to the development of machines and heat engines. Scientific principles of electricity and magnetism led to the development of the lamp, the telegraph, the telephone, the Internet, the radio, stereo systems, the television, radar, and so on.

Modern physics led to the development of semi-conductor devices and large-scale integrated circuit chips, lasers, energy conversion technology (e.g., fission, fusion, solar energy, and fuel cells), X-rays, and other diagnostic technology. Feedback theory led to development of sophisticated control systems for industrial processes, space flights, and the environment. Computer science applications led to the revolution in modern computer technology. Discoveries of new chemical reactions led to the development of new chemicals and medicines. Developments in genetic science led to the development of new treatments in healthcare technology.

Indeed, we marvel at the progress in communication, control, computers, and healthcare technology. Developments in these four areas of technology alone have increased our material comforts beyond our wildest dreams. To sum up,

A. **Communication** – utilizes the concepts of information science, electrical signals, and electromagnetic fields in telephones, radios, stereo systems, TVs, radar, etc.

B. Control – utilizes the concept of feedback for control over climate and environment, energy transformation and utilization, transportation, and industrial production.

C. Computing – utilizes electronics, mathematics, and information science for computing and data processing, control, and decision-making.

D. Healthcare – utilizes various concepts in modern physics and chemistry for health-related technology

Developments in the above-mentioned fields of technology have been phenomenal in the past century. During this very brief period compared to our existence on this planet, we have indeed made great strides. Gone are the smoke signals for communication; lack of control over temperature in summer and winter; lack of food and material goods, and lack of energy for lighting, heating, and transportation, lack of medicines, diagnostic tools, and technology to treat several diseases.

Undoubtedly, because of the technology, we are materially more comfortable today than we were a hundred years ago. However, we have just started. Despite tremendous progress, we have barely scratched the surface. The future of technology looks very promising as we march on. As soon as someone predicts the end of the road, a new scientific discovery comes along, opening a new road for technology.

1.1 Technology based on Simple Mechanics

Humanity has always looked for help to perform menial tasks requiring physical effort. First, we built tools from stones, spears, and bows and arrows for hunting and fighting. Then, we domesticated animals and built some gadgets for agriculture and wheels and carriages for transportation. We also learned to use levers and pulleys to lift huge stones for building shelters and pyramids.

Until the seventeenth century, we built a few gadgets, but it was Newton's scientific discoveries that brought about a remarkable turn of events. Newtonian mechanics ushered in powerful machines that could build other machines and utilize energy locked in various natural resources. This enabled us to amplify our mechanical efforts. The development of the steam engine by Watts followed his chance observation of a teakettle's lid chattering due to steam pressure from the boiling water. Gradually, we developed an understanding of heat energy and thermodynamics.

The conversion of various forms of energy to mechanical form brought in new factories and ushered in an industrial revolution spanning entire continents. It ushered in an industrial age as the Western countries moved from being agriculture-based to industrial-based societies. Our world suddenly became much more comfortable to live in. We built railroads, trains, cars, roads, and airplanes to travel faster and farther. We also started developing technology to control our environment, temperatures, and so on.

Imagine life before and during Newton's time. It was short-lived and miserable, with the average life expectancy being less than thirty years. People struggled just to survive. Starting at early dawn, most people worked all day in fields during winter and summer, returned home, ate, and then slept through the dark nights only to wake up to the same routine the next morning. There were no trains, planes, or cars to go anywhere. We had only animal hides and fire to prevent us from freezing in winter and no protection from hot summers except running to shade. There were no electric lights at night and no entertainment except eating, drinking, and sexual pleasures if one could avoid famine and diseases and stay healthy. It must have been a miserable life as far as material comforts were concerned. Technology has indeed changed our lives in the past two centuries!

Ironically, the abundance of food and mechanical help from machines started us on a course to make us lazy and obese. We now eat more, ordering super-sized food packages at restaurants, and do little walking and physical work – thanks to technology. However, we should not blame technology for this. We have choices. We need the discipline to exercise these choices intelligently. By controlling our diets and exercising regularly, for example, we can have the best of both the worlds, namely, we can enjoy the fruits of technology and live a good life.

1.2 Technology based on Electricity

In the mid-1800s, as new discoveries were being made in electricity and magnetism, inventors were trying to develop an affordable electrical home lighting device. Thomas Edison invented the electric bulb just around 1878. The basic science behind the invention of an electric bulb is simple, namely, conversion of energy from electricity to light. The electrical current flow heats up a tungsten metal filament, which starts radiating light.

This easy-to-use technology was such an improvement over the old ways that the world never looked back. Within twenty-five years, millions of people around the world had electrical lighting in their homes. The electric lights lit up the entire planet.

By the way, when Thomas Edison invented the electric bulb, those present questioned its usefulness. An Oxford professor predicted the end of electric light soon after an exhibition in Paris in 1978. Edison persisted, nevertheless, and did reap the benefits from his invention. Thomas Edison was very fond of saying that a genius is 1 percent inspiration and 99 percent perspiration.

Around the same time the electric bulb was invented, electricity was being put to use in many other ways. The telegraph signal was already being transmitted, and the telephone was being invented, which ushered in the era of communication technology. The progress in the field of communication technology during the past century has been phenomenal.

These days, we can communicate a signal from any spot to any other spot, and computers can communicate with each other worldwide and across space over long distances. It has been quite an exciting journey, starting with telegraphy and Morse code, when Morse, in 1844, telegraphed the words "What hath God wrought?" from Washington to Baltimore on wires.

On March 6, 1876, the telephone was invented when Bell, from one room, called his assistant in another room. "Come here, Watson; I want you." Watson heard it through a receiver connected to the transmitter that Bell had designed, and the Bell Telephone Company (later AT&T) was born, which grew to be the largest telephone company in the world.

The basic science behind telegraphy and telephony is simple. In telegraphy, we transmit a code built from a sequence of zeroes and ones by repeatedly switching an electric circuit off and on. This coded electrical signal is received on the receiving end and decoded to extract the coded message.

In telephony, we convert our speech to electrical signal with the help of a simple transducer inside the telephone called a microphone, which converts sound pressure waves to corresponding variations in an electrical signal. This signal, in a modified form, is transmitted and converted back to speech at the receiving end. A simple transducer inside a telephone receiver converts electrical signal back to mechanical vibrations corresponding to the transmitted speech.

This journey bringing new inventions has continued as we moved on to wireless radio, stereo systems, television, satellite communication, cell phones, fiber optics, and laser and computer communications.

Wireless communication owes its origin to the propagation of electromagnetic waves. It evolved after Maxwell developed his equations in 1876, which give a complete description of electromagnetic phenomena. These equations predicted the propagation of electromagnetic waves. Hertz showed that these waves indeed exist. The next step followed naturally. The electrical signals corresponding to voice (e.g., generated in telephony) were sent riding on the carrier electromagnetic waves propagating in space.

Marconi sent a wireless signal from St. John's, Newfoundland, across the Atlantic Ocean. It happened just over a century ago, and the modern-day, wireless communication systems were born. Advances in wireless communication have been truly astounding. This rapid pace has led to serious challenges for the planners of radio spectrum, who allocate different frequency bands to radio and TV stations, military, police and cell phones.

- Technology Development Cycle

While discussing communication technology, let us discuss an important point for the development of technology. We are often impatient and ask why it took such a long time to develop the present-day communications after Maxwell's scientific discovery.

This fact illustrates a very important point regarding the development of any technology. It takes considerable time to bring scientific ideas to fruition, since the ideas have to pass through several stages before they turn into mature technology. The technology development cycle, like most operations, involves certain basic steps, as shown below.

A. Design– Plan to develop a technology

B. Review– Review of plan by entrepreneurs and engineers

C. Execute – Production of technology

D. Evaluate– Quality control, marketing, customer response

E. Modify– Implementing required changes

For example, for communication technology, demonstrating that a signal can be transmitted to a distant place (telecommunication) is only the beginning. This demonstration requires a basic understanding of electrical science and building a crude transmitter and receiver from crude electronic devices.

Then come questions such as, how far can we communicate? How many signals can we send on a single channel? How can we improve the quality of reception? Science and technology must join hands to find practical answers to these questions. We also need to develop appropriate science materials and build the engineering bridge needed to design communication systems.

Next, the search begins for the entrepreneurs to entice them to commit resources for further development. They review the plan and decide to take a calculated risk by providing the resources to develop marketable communication systems.

As plans are executed and technology develops, marketing groups spring into action to develop demand by demonstrating its need and value and monitor customers' reactions and market needs. Based on evaluations of consumers' reactions, the course is further charted for the development of the technology. The process goes on as technology matures and more companies jump in the arena.

By now, the government has already stepped in. For example, during the evolution of communication technology, the government steps in to formulate rules and regulations governing communication and to settle any disputes. It also allocates different bands of frequency in the electromagnetic frequency spectrum for public use and collects revenue through the sale of different bands. The government also collects part of the revenues as taxes from the industry, some of which are funneled back into the research projects funded by various government agencies.

That is how any discovery in science evolves into a mature technology. Further refinement of technology continues. For example, as communication technology matures, communication engineers get more ambitious and continue to improve technology. They want to cram more signals on to a carrier within the allocated frequency band, communicate them simultaneously over longer distances without losing signal strength, and receive noise-free signals at the receiving end.

- Communication System Evolution

Returning to the evolution of communication technology, let us first clearly understand the basic components of a communication system, as shown in Fig. 1.1.

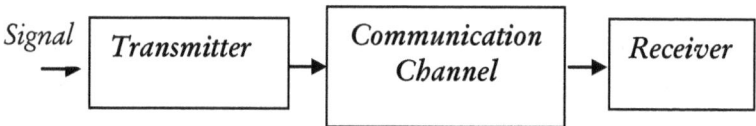

Fig. 1.1 – Basic Communication System

- **Signal** - The message or information we want to transmit. Anything else is considered noise.
- **Transmitter** – Converts signals into electrical signal, mixes or modulates it with electromagnetic carrier wave, and transmits it over the communication channel.
- **Communication Channel** – The electromagnetic waves travel to the receiving end over the communication channel. It can be in the form of atmosphere, free space, satellite in space for line of sight transmission, and/or fiber optic cables, etc. The allocated bandwidth and disturbance in a channel limit a channel's capacity.
- **Receiver** – Receives and recovers the original signal. It de-modulates or separates the signal at the

receiving end. The fidelity of the receiver or quality of the received signal depends on the modulation schemes, the transmitter equipment, the communication channel, the demodulation schemes, and the receiver equipment.

The following important developments had to take place in science and technology after demonstrating the feasibility of communication (with or without wires).

- Science Related to Signals

Signal representation theory represents analog deterministic signals as a summation of sine and cosine waves of different frequencies, using Fourier series for periodic signals and Fourier transforms for non-periodic signals. Similarly, random signal representation was developed in terms of power spectral density as a function of frequency, which is the Fourier transform of the auto-correlation (i.e., measure of how well a signal self co-relates). In addition, discrete and digital theories for signal representation were developed.

A signal contains information. How do we measure information content? Shannon Information Theory, developed around 1930, answers this question. The basic idea is that information is a measure of entropy. The more chaos or random variation in a signal (or the less likely it is to be anticipated), the more information it contains. A constant tone or a single frequency signal carries little information.

The faster a signal varies in time, the more frequencies it would need for representation in terms of Fourier series or transform. In other words, it has a broader frequency spectrum or bandwidth. For example, an intelligible speech signal conveyed over telephone lines contains frequencies only up to 400 kHz. It has a bandwidth of around 400 kHz. A high-quality music signal would have a bandwidth of around 16 kHz. A video or TV signal has a bandwidth of several MHz.

- Science Related to Receivers and Transmitters

A transmitter modulates the audio or video signals before transmitting them over the electromagnetic carrier waves in different frequency bands, allocated by a government agency. Modulation theories describe how to make the signal ride on very high frequency electromagnetic carrier waves, which would carry the signal and propagate over the communication channel.

The theory starts with analog signal communication, first altering the amplitude of a carrier sine wave in accordance with a signal (i.e., amplitude modulation [AM]), and then altering a carrier's frequency with a signal (i.e., frequency modulation [FM]). Frequency-division multiplexing techniques were also developed to accommodate more than one signal on the same channel within the allocated bandwidth.

Digital sampling, digital signal processing, and filtering theories ushered in the era of digital communication. Digital communication involves sampling a signal, quantizing it, and coding it into a stream of binary bit sequences of ones and zeroes. We then divide the high-frequency electromagnetic carrier wave in time, and the signal bits are pulse-code modulated in time. Since a digital signal is a coded binary signal in a discrete packet of ones and zeroes, it is easier to detect and process this signal at the receiving end.

We can easily clean up the digital signal after transmission, and identify the exact transmitted sequence of ones and zeroes. Error-correcting schemes are also utilized to ensure the recovery of the transmitted sequence of ones and zeroes. We can make the received signal virtually error free by detecting the presence or absence of pulses corresponding to one and zero.

In the analog signal, detection of continuous amplitude levels is more difficult because the signal is always corrupted with noise in various stages of communication. However, in digital communication, a receiver has to recognize essentially whether the incoming signal bit is absent (0) or present (1) at a particular instant, instead of the actual signal value, as in analog communication. Digital communication has a far superior signal-to-noise ratio, which implies that we can recover the signal at the receiving end even in the presence of excessive noise.

The evolution of communication system transmitters and receivers parallels the evolution of communication science and technology. Earlier in the age of analog communication, we commonly used the amplitude modulation (AM) and frequency modulation (FM) techniques by varying the amplitude or the frequency of the carrier in accordance with the signal. Nowadays, in the age of digital communication, we sample the signals in time, quantize and digitize in binary form. Then these signals are pulse code modulated, and time-division multiplexed on a carrier.

To recover a good quality signal at the receiving end, we have new digital modulation schemes these days at the transmitting end, which improve the signal-to-noise ratio. We also use combinations of various modulation techniques to improve the quality of the received signals, and send many such signals simultaneously over the allocated bandwidth in the Radio Frequency Spectrum.

In the United States, the Federal Communication Commission (FCC) regulates and allocates the bands in the radio spectrum. Fig. 1.2 shows a typical spectrum allocation chart. Headquartered in Geneva,

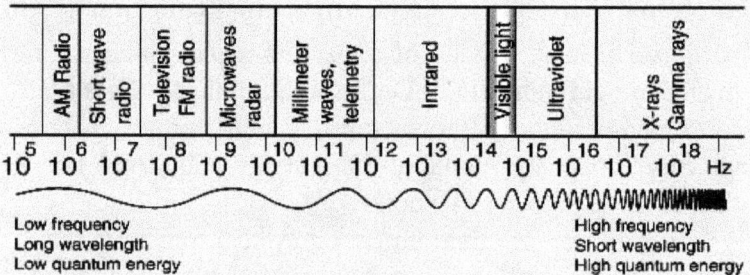

Fig. 1.2 – Radio Frequency Spectrum

The International Union coordinates the world's communication policies at the international level.

- Science Related to Communication Channels

Concepts of channel capacity, channel bandwidth, channel noise, and so on had to be developed to take full advantage of the communication revolution. We stated already that Shannon, in 1948, in a paper in the Bell Technical Journal, unified all phenomena related to classical information processing by quantifying the information content produced by an information source.

He defined the information as the minimum number of bits needed to store reliably the output of the source. His expression, known as Shannon entropy, plays a central role in data compression, information transmission over noisy channels, and analyzing stock market behavior or behavior of any system that processes information.

As the carrier frequencies increased from radio waves to microwaves and light waves and the radio spectrum allocated for various types of communication broadened, the communication channel parameters kept changing to keep pace with the changing needs.

Fiber optics, lasers, and other modern devices have played a very important role in realizing the modern era of communication. In principle, photons can transmit, manipulate, and store information more efficiently than electrons can. Thus, optical fiber cables have replaced the copper wires that were used for data transmission for more than a century.

Regarding the development of communication technology, we must reemphasize the point that the technology usually lags and does not keep pace with science. Technology needs time to develop new devices that are needed to turn a scientific concept into a practical application.

In this regard, let me narrate an interesting incidence. During the summer of 1975, Dr. S. P. Chan—a friend from my graduate study days at the University of Illinois (1960-63)—invited me to teach an advanced digital-signal processing course at the University of Santa Clara. The students were mostly research scientists and engineers from Hewlett-Packard.

These students asked about devices like VCRs and the potential of digital communication. My answer was that communication science and the digital signal processing theories were well developed, and it was only a matter of time before technology caught up with science. Then, we must watch out, because it is difficult to see the limits.

Soon thereafter, the market was flooded with VCRs, compact disc (CD) players, and analog camcorders. Now, the market is flooded with digital video recorders (DVRs), digital video disc (DVDs), cell phones, digital TVs, and so on.

- Internet Technology

It is rather ironic that scientists and technologists working for the defense establishments developed many technologies related to communication, control, and computers. The Internet evolved around 1977 from the Pentagon's ARPANET (Advanced Research Projects Agency), and they developed it to win a possible nuclear war. E-mail came into existence to meet the need to communicate during and after such a nuclear war. During the process of this development, it linked up the scientists and a few universities in the United States. It became quite successful in its mission, and it was made available to the public through the National Science Foundation (NSF) in 1990. E-mail has become the most common method of instant communication.

The end of the communication technology revolution is not yet in sight. The high speed of the Internet and the rest of the tightly connected, lightning-fast communications network owes its existence to a 1966 innovation—fiber optics—in which we transmit light signals over long distances via hair-thin glass fibers. Charles Kao at Standard Telecommunications Laboratory in Harlow, England, purified glass to unsurpassed transparency.

In 1988, the first fiber optic cable was laid between Europe and North America. Kao was the joint recipient of the 2009 $1.4 million Nobel Prize for physics for the accomplishment. The remaining 2009 Nobel Prize for physics was shared between Willard Boyle and George Smith for their invention of the eye of digital cameras, the charged coupled device, or CCD. A CCD is a sensor that transforms light into the pixels, or dots of color, that are the building blocks of digital images.

The Internet brought a revolution in communication. On the Internet, computers and servers link in a giant network—the "mother" of all networks. The Internet revolution knows no bounds. It is spreading like fire. There has been no looking back since the Internet exploded into the public domain.

Since its advent, it has been growing daily and rapidly, and it is now as widespread as the worldwide telephone system, with billions of connections. In the United States alone, it has grown from usage in 18 percent of households to 65 percent from 1997 to 2007. More recent data shows that the Internet usage is growing like wild fire. It has brought the people on this planet togather like one family. There is no end in sight for this phenomenon.

The amount of information already stored and accessible on the Internet is phenomenal. We can now access almost any information we desire on various worldwide web (www) sites. Recently, the British developer of the worldwide web (www), Sir Tim Berners-Lee, said that he is worried about the use of www to spread misinformation and "undemocratic" forces.

We are already realizing the full economic and commercial potential of the Internet. The web has transformed the way many people work, play, and do business. It has become a common and convenient way of shopping, and almost all businesses have set up their own websites. Banking online has also become quite popular. Soon, we will have all the knowledge and collective wisdom of humanity, along with a lot of useless information, available on the Internet.

Industry related to the Internet has mushroomed. A company like Google with powerful search engine is making billions of dollars. Many other companies are also flourishing as they develop many applications for the Internet.

Internet technology has already ushered in a new era with amazing progress. Social networking applications like Facebook have added new dimensions to connectivity of friends, families, and professionals. Availability of Internet access on mobile phones and devices like iPAD has given a new dimension to this technology. As the technology matures, we hope to see more stability, drastic improvements in the Internet technology, and in the performance of the related industry.

- GPS and Mobile Technology

Another marvelous application of communications is geo-positioning (GPS) system technology. Users of GPS technology can get accurate location information across the globe, which they can interpret to provide speed and distance to a destination.

The US Defense Department developed this technology and made it available for public use in 2000. The US military originally built NAVISTAR, launched the first satellite in 1978, and made the GPS system operational by the early 1990s. The defense industry has made extensive use of the GPS receivers in smart bombs and missiles, which can guide these weapons to hit the designated targets.

The basis of GPS is a constellation of twenty-four solar-powered orbital satellites about twelve thousand miles above the earth. A GPS device—a tiny receiver on the ground—seeks tracking signals from at least three satellites and then interpolates the data to establish latitude and longitude. If a device can pick up four or more signals, it can also determine the user's altitude.

The GPS device can tell you where you are on the earth within a few feet by comparing timing signals broadcast by these GPS satellites. Advanced systems known as differential GPS and the Wide Area Augmentation System incorporate the use of stationary ground stations to interpolate signals and can offer accuracy of 1 to 3 meters or better. The satellites have diagnostic systems, and they stop sending position data if they are malfunctioning.

For accurate distance measurements, the timing signals have to be accurate to around one second in ten thousand years, achieved by using atomic clocks regulated by the frequency of microwaves that correspond to the quantum transition of an atom. The Global Positioning System also requires corrections to time, based on the relativity's effect on time. Each GPS satellite moving around at about 14,000km/hr would imply that the onboard atomic clock lags behind the earth's clock by about seven microseconds on a daily basis.

Furthermore, the gravitational pull on the GPS satellite, placed at a height of about 20,000 kilometers, is about one quarter of the gravitational pull on the ground, depending on the orbit. The resulting relativistic effect makes the onboard clock run faster by about 45 microseconds per day. Therefore, if the onboard clocks were not turned back by 38 microseconds every day, it would cause an error of 11 kilometers per day that would keep building up.

Some people, while launching the first GPS satellite, did not believe in relativistic effects, but made such a provision half-heartedly. They soon found out that we need the provision, and they turned the switch on.

Most large ships and aircraft have these devices, and the FAA has incorporated GPS as an essential part of instrument flight. The Global Positioning Systems (GPS) are now common in many cars, and transit companies use them to track trucks and buses. Several cheaper models in the range of couple of hundred dollars are available in stores.

Hand-held devices are being used by everyone from hikers to land surveyors. For a few hundred dollars, one can buy a GPS device that speaks aloud—not just relaying the directions to where you want to go, but also describing the sights when you reach them.

Isaac Daniel, an engineer, recently embedded a GPS chip into a line of sneakers that can locate the person wearing them anywhere. He wishes his eight-year-old son had been wearing them when he got a call from his school in 2002 saying the boy was missing. It's the latest implementation of satellite-based navigation into everyday life—technology that can be found in everything from cell phones that help keep kids away from sexual predators to fitness watches that track heart rate and distance.

The development of the cell or mobile phone has been another revolutionary aspect of communication technology. One can now call any person with a cell phone from any corner of the world. In developing countries like India and China, the advent of mobile phones has brought a phone within the reach of even the poorest persons in the country. It has been a truly revolutionar development.

With new generation of mobile phones, one can receive or send email, connect to a computer, access any information on the Internet (including real-time stock market reports), take high-resolution photographs, and control remotely located appliances. In fact, soon, we should be able to do most tasks on a mobile phone.

Chapter 2
Energy & Modern Technology

2.1 Introduction

We continue our journey and first visit various milestones in modern science that led to the development of new technology. We start with energy conversion technology. We discuss various resources available for energy conversion. These sources include chemical, thermal, hydro, nuclear, solar, hydrogen fuel cell, wind, tidal, geothermal, and biomass. Then, we visit other areas of technology, which are based on modern science.

2.2 Energy Conversion Technologies

Before we focus on the currently available technologies for energy conversion, let us first discuss all the forms of energy conversion. We can store energy in many forms and transform it from one form to another. Although we do not currently use some of these techniques, they are potentially important as we look for new renewable resources. According to some estimates, the world is going to run out of oil and gas in about fifty years, and we must develop new technologies to maintain civilization.

Table 2.1 lists different forms of energy conversion and various transducers used for the conversion.

ENERGY	Mechanical	Sound	Electric	Magnetic	Heat	Light	Chemical
Mech.	X	Drums	Electric Generator	Magnetize Iron	Friction Devices	Friction Devices	Electrolysis
Sound	Telephone Transmitter	X	Micro-phone				
Electric	Electric Motor	Loud-speaker	X	Electro-Magnet	Electric Heater	Light Bulb	
Magnetic	Magnetic-levitation Train			X			
Heat	Steam Turbine		Heat Two Metals Junctions		X		
Light	Laser Gun		Solar Cell		Solar Heater	X	Plants
Chemical	Heat Engine		Batteries		Gas Heater	Phosphor-escence.	X

Table 2.1 – Energy transformation and transducers

Table 2.1 includes, directly or indirectly, electrical energy, solar energy, chemical energy, wind energy, geothermal energy, tidal energy, and so on. However, it does not include some important forms of energy. These include gravity, nuclear energy, "dark" energy, or so-called "vacuum" energy.

Electrical energy is used for many applications. There are two types of technology for generating electrical energy. In indirect conversion technology, we usually first transform the given energy into a mechanical form. Then, we convert this mechanical energy into electrical energy by rotating the rotor of an electric generator. This results in a relative motion between magnetic and electric fields, generating electricity. Examples of such technology are chemical, thermal, hydroelectric, and nuclear power stations.

In direct conversion technology, we directly convert the given energy into electrical energy. Examples of such technology are solar cells, hydrogen fuel cells, and so on.

In electrochemistry, through fuel cell technology, we can strip electrons from hydrogen traveling through a membrane in a cell and produce electricity. We can convert the resulting electric current into mechanical energy using an electric motor, which in turn can run a vehicle or do other work. Instead of a hydrogen fuel cell, we can use electric batteries, which convert chemical energy into electrical through electrolysis.

We discuss next some of the alterative conversion technolgies, such as, wind, tidal, biomass. As regards the gravitational energy, we have not been able to utilize this energy, except in few instances, such as in hydraulic power stations. In such power systems, we convert the potential energy (due to gravity) of water stored at a certain height into kinetic and then electrical energy. However, gravitational energy is extremely important. The motion of stellar objects, stars, and galaxies revolving in their orbits mostly tie up and engage gravity.

With regard to nuclear energy, we first misused nuclear energy by making thermonuclear bombs. We then harnessed nuclear energy and built nuclear power stations. Based on nuclear fission in atoms, we converted the released nuclear energy into electrical energy. We are still trying to harness the nuclear fusion energy, and w do not yet understand "vacuum" or "dark" energy.

No energy conversion is 100 percent efficient, since the conversion process entails inevitable losses. However, some conversion techniques are more efficient. For example, fuel cell conversion efficiency is around 55 percent. Fuel cell technology is more efficient, and it is pollution-free, since the only byproduct is water. A four-stroke internal combustion engine has efficiency up to 30 percent.

It is also interesting to note that we can improve the overall conversion efficiency through multi-tasking. For example, co-generation technology has evolved recently and involves cascading several conversion techniques to extract as much useful work from a given form of energy as possible. A simple example is utilizing the heat energy remaining in the spent steam from the turbine exhaust to heat water that can be used to heat buildings.

We now discuss the following available technologies.

A. Chemical and Thermal
B. Hydroelectric
C. Solar
D. Nuclear fission and fusion
E. Hydrogen-fuel cell
F. Wind
G. Tidal
H. Biomass

- Chemical and Thermal

A form of energy that we use extensively is the chemical energy stored in fossil fuels. We can release this energy and convert it into a more useful form. For example, we can convert it into mechanical energy by igniting the hydrocarbons in an internal combustion engine, which can move a piston up and down, providing propulsion. In a four-stroke engine, a piston connected to a crankshaft can run a car or fly an airplane. We also burn fuel in jet engine planes and use reaction force to propel a plane.

We use coal or gas in a thermal station to generate steam and convert its heat to mechanical energy through a steam turbine. We then convert the resulting mechanical kinetic energy into electricity by imparting relative motion between wires and magnets in an electric generator. Instead of coal or gas, we also convert the potential energy stored in water dams to generate electricity.

The United States of America imports over twenty million barrels of oil and vast amounts of natural gas in addition to its own production to meet part of its energy demand. It also generates almost half of its energy needs from its vast coal reserves. Unfortunately, the amount of CO_2 released in these processes is having a profound effect on the environment. According to some scientists, if we do not come up with clean coal technology in the next few decades, it might be too late.

The United States built one such plant at the cost of several billion dollars in the seventies to remove CO_2 from natural gas and send it underground instead of releasing it into the atmosphere. The cost of such plants for clean coal technology could easily exceed a trillion dollars. Furthermore, if we inject such huge amounts of CO_2 into the ground, we need to understand what happens to this gas. If it percolates out to the surface, it would still affect the environment. However, if we want to continue to use coal, we must quickly develop some form of clean coal technology.

We also use the thermal energy stored underground for power generation. Geothermal energy is the heat stored inside the earth. We utilize this energy by transferring the heat from the inside to the surface of the earth. Geothermal energy is relatively economical, but not readily available or obtainable.

- Hydroelectric Technology

In hydroelectric power systems, we convert the potential energy (due to gravity) of water stored at a certain height into kinetic and then electrical energy. Hydroelectric power stations employ huge water turbines to drive electric generators. As we release the water stored behind a dam, its kinetic energy rotates the turbine blades and generates electricity. Hydroelectric power systems require enormous capital expenditure upfront. However, the maintenance costs are relatively low, and they provide power quite cheaply. In the United States, approximately 180,000 MW of hydroelectric power potential is available, and we currently harness only about a third of the available potential.

- Solar Technology

Because of the high cost and environmental problems associated with the use of energy from fossil fuels, scientists are considering the use of solar energy. There is plenty of solar energy. For example, the solar energy received from the sun in 48 minutes is enough to meet the world's energy demands for the entire year. According to an article by Zweibel, Mason, and Fthenakis in Scientific American in January 2008, if we could convert in 2006 only 2.5% of the solar radiation in the southwest region of the United States, it would meet the energy consumption in the USA for the entire year. The grand plan described in the Scientific American article could generate 3,000 gigawatts (GW) of power. According to this plan, with the 14 percent efficiency of solar devices, the installed cost would be around $1.20/W, and one could sell this electricity at 5 cents per KWh by 2050.

The solar energy received from the sun is about thirty-five thousand times the energy we use on this planet. It comes from the radiation produced by nuclear fusion reactions deep in the sun's core. One square kilometer at the outer edge of our atmosphere receives approximately 1,400 megawatts of solar power every minute. The amount of light that reaches any point on the ground depends on the time of day. Only about half of the amount from the outer edge of the atmosphere reaches the earth's surface.

Regarding the solar energy conversion directly to electricity, Einstein described the basic principle for the photovoltaic phenomenon in 1905. Einstein's basic idea was quite simple. Suppose we shine light on a metal, and the frequency of the incident light and the associated energy is high enough. Then it would release free electrons from the metal and generate electrical energy in the form of free electrons.

If the energy E of the incident light of frequency, f, ($E = hf$, where h is the Planck Constant) exceeds the characteristic energy, called the work function of the particular metal, ($\phi = h f_t$), corresponding to the threshold frequency (f_t). Then the metal atoms absorb the light quanta, and the energy quanta make a peripheral electron free, no longer bound to the atom. Thus, we convert the light energy into the electrical energy in the form of free electrons. Solar cells, used in many applications, convert light from the sun into electrical energy using this principle.

Some of the advantages of solar energy are as follows:

1) Solar energy is freely available, renewable, and pollution-free.
2) It integrates well with other clean systems.
3) The cost for producing electricity from solar energy has dropped significantly. The modules of thin film made of cadmium telluride have achieved a conversion efficiency of almost 10 percent.

Some of the disadvantages are as follows:

1) The electricity produced is more expensive.
2) The conversion efficiency of a solar cell is only around 8 percent.
3) It cannot be the only system in cloudy places.
4) Energy has to be stored in batteries, hydrogen, water, or other matter.
5) (Excessive power produced on sunny days must be stored for use during cloudy days and nighttime. Battery storage would be inefficient and costly. Compressed air-energy storage in pressurized caverns and concentrating sunlight falling on a pipe filled with fluid and running this hot fluid through heat exchangers are viable alternatives.)

6) We might have to replace the current alternating current (AC) transmission system with direct current (DC) transmission systems, which, by the way, waste less energy.

- Fission Technology

We can release the binding energy from atoms of certain elements such as uranium-235 through nuclear fission and through fusion by combining two nuclei of hydrogen to form a nucleus of a heavier element. We use the released energy in these processes to explode atom and hydrogen bombs, or use it by converting it into useful forms of energy, heat, electricity, and so on.

We generate this power by splitting the atom. In science, we learned that an atom's nucleus has protons and neutrons that are held together by a strong force. The mass of the nucleus is always less than the combined mass of the protons and neutrons in the nucleus. This difference in mass Δm in terms of its equivalent energy ΔE is the nuclear binding energy that holds the nucleus together. Nuclear fission is the process by which a nucleus breaks down into two or more smaller nuclei, releasing this binding energy.

The energy ΔE released equals $\Delta m.c^2$ according to the Einstein's famous equation $E=mc^2$. The Δm denotes the difference in mass between the original nucleus and the combined mass of the two smaller nuclei, and c is the speed of light. For example, the uranium U-235 nucleus is unstable, and when a neutron from the original U-235 nucleus affects another nucleus, it causes fission, starting a chain reaction with further collisions. A critical mass of U-235 can sustain this chain reaction, releasing a tremendous amount of energy. The amount of energy thus released is huge because of the c^2 term where c is the speed of light (~$3x10^8$m/s).

This aspect of modern physics forms the basis for the development of technology for releasing uncontrolled energy through atom bombs. We saw the destructive power of an atom bomb (small compared to recently developed bombs) dropped on Hiroshima during World War II, where only 1 percent of two pounds of uranium was converted into energy. Later, the technology to control the chain reaction in a nuclear reactor then evolved to harness this nuclear fission energy.

It is interesting to note that about 360 nuclear fission plants generate about 24 percent of the industrialized world's power. More countries are building nuclear fission plants to generate electrical power. However, the potential for disaster, such as at Chernobyl and Three Mile Island, and the failure to develop technology for the safe disposal of generated nuclear waste, remain problems of concern.

A one thousand-megawatt nuclear fission plant produces about thirty tons of nuclear waste. When we add up the amount from over one hundred such commercial plants in the United States alone, the resulting nuclear waste will be extremely dangerous to life for millions of years. States are hesitant to provide even land for deep burial of such wastes. It is said that the cleanup at seventeen disintegrating nuclear weapons sites in the United States may cost up to five hundred billion dollars.

- Fusion Technology

In the nuclear fusion process in nuclear physics, two nuclei collide and join to form a heavier nucleus. For example, if we can supply enough energy to push two nuclei of hydrogen together, it can fuse these two nuclei. This could convert them into a helium nucleus through the fusion reaction, releasing tremendous amounts of energy because of the resulting mass difference.

This aspect of modern physics forms the basis for the development of technology for releasing uncontrolled energy through hydrogen bombs. When we go out on a sunny day, we access fusion energy: The energy from the sun comes from nuclear fusion, where the hydrogen atoms crush into helium atoms and convert the small amount of mass into energy.

A fusion reactor would extract power from energy released in nuclear fusion. The neutrons released in the process heat the water, turning it into steam. Unlike nuclear fission, nuclear fusion does not create high-level nuclear waste. The only waste would be any radioactive hydrogen that might escape the steel hull that may become radioactive over time. Nuclear fusion plants will not melt down or go super-critical. It is projected that 10-20 percent of the world's energy could come from fusion by the end of the century.

The problems with designing safe fusion reactors are easy to understand, but difficult to solve. Hydrogen nuclei are protons, and they repel each other because of the positive charge. One must heat these nuclei to a very high temperature (to one hundred million degrees) to overcome this repulsive force, slam the nuclei together so they can stick together, and fuse. In a hydrogen bomb, we achieve such temperatures by first detonating an atom bomb.

In stars such as the sun, the intense gravitational field at the core compresses the hydrogen gas in the stars (called gravitational confinement) and heats it to the required temperatures, fusing the hydrogen nuclei. Obviously, one cannot use such gravitational confinement on Earth. Instead, intense magnetic fields or powerful laser beams confine the plasma in several designs. However, several problems related to confinement prevent controlled power from nuclear fusion from becoming a reality.

So far, we have not been able to develop technology to control the fusion reaction and design fusion reactors and power stations. However, fusion technology has the potential to provide energy for the whole world from the supply of seawater. For decades, scientists have been trying to discover how to harness the fusion reaction to generate electrical power. They would like to reach a milestone, the "break-even" point, at which a controlled fusion reaction produces more energy than it consumes. Research trying to achieve this goal is progressing on the following three main fronts.

Laser fusion: The US National Ignition Facility, a $3.5 billion laser research site, has been built at California's Lawrence Livermore National Laboratory (NIF). It is designed to produce fusion power on a small scale by aiming one hundred and ninety-two laser beams simultaneously at a hydrogen target the size of a pencil eraser for a burst lasting just a few billionths of a second.

In the NIF approach, called inertial confinement fusion, the target is a centimeter-scale cylinder of gold called a hohlraum. It contains a tiny pellet of fuel made from an isotope of hydrogen called deuterium. NIF recently swept aside a significant potential concern to the process, namely, that the plasma, a roiling soup of charged particles, would interrupt the target's ability to absorb the laser's energy and funnel it uniformly into the fuel, compressing it and causing ignition.

Siegfried Glenzer, the NIF plasma scientist, led a team in 2010 to test this theory, putting to rest this concern and smashing the record for the highest energy from a laser by a factor of twenty by using all hundred and ninety-two of its laser beams.

Inertial electrostatic fusion: Another approach is known as inertial electrostatic confinement fusion, or Polywell fusion. This method, proposed by the late physicist Robert Bussard, involves designing a high-voltage cage, in which atomic nuclei slam into each other at high speeds, sparking fusion.

Magnetic fusion: So far, most of the fusion research is focused on magnetic containment of fusion plasma, usually within a doughnut-shaped chamber, known as a tokamak. The important current research for magnetic confinement fusion is the ITER project, headquartered in southern France.

Research continues in this area, funded by various governments throughout the world. For example, the United States spends around $400 million and Japan spends 40 percent more than that on the research and development of fusion technology. In February of 2005, the United States decided to become a partner again in the five billion-dollar International Thermonuclear Experimental Reactor (ITER) fusion experiment.

This research effort, conceived in 1986, would pave the way for fusion power. It will use a doughnut-shaped magnetic bottle to confine super-hot hydrogen plasma and induce it to undergo nuclear fusion. The seven-party consortium—which includes the European Union, the US, Japan, China, Russia, and others—recently agreed to build ITER in Cucaracha, in the southern French region of Provence—a ten billion-Euro project. If all goes according to plan, officials hope to set up a demonstration power plant in Cucaracha by 2040.

One kilogram of fusion fuel would produce the same amount of energy as ten million kilograms of fossil fuel. The development of technology to obtain huge amounts of energy through fission and fusion brings up a very important point. Science points out the concept through which we can produce this tremendous energy. Technology helps us produce it. However, it is up to humanity to use or misuse this energy. On one hand, we can misuse this energy in an uncontrolled manner to explode nuclear bombs and wipe out civilization on our planet.

The possibility of cold fusion is also receiving some attention. The American Chemical Society organized a special session in 2009 to discuss this possibility. The principle of cold fusion is different from the other fusion production mechanisms, which utilize enormous lasers or magnetic chambers that contain hot gas.

In 1989, Pons and Fleischmann performed a simple experiment at room temperature, and passed a current through an electrolytic cell. They observed heat rise in the cell, which seemed to indicate that nuclear fusion was producing power within it. However, numerous attempts to repeat the experiment around the world failed to establish cold fusion as a reality.

The control of any process to produce nuclear fusion energy is critical. We shall shortly discuss truly revolutionary advances in control systems. For now, let us continue our journey by visiting a few other energy conversion technologies based on modern physics.

- Hydrogen Fuel Cell & Battery Technology

Hydrogen energy is a potential primary source of fuel for automobiles, heating buildings, and generating electricity. An electric car can store its energy, or it may generate energy using a fuel cell or generator. A fuel cell is a special battery that combines hydrogen with oxygen in a chemical reaction that generates electricity and water vapor.

A fuel cell essentially converts hydrogen gas into electricity. It does so by stripping electrons from hydrogen, traveling through a membrane in the cell. One can construct a proton-exchange membrane (PEM) fuel cell by combining two porous electrodes (i.e., an anode and a cathode) separated by a polymer membrane electrolyte that permits only protons to pass.

Catalysts coat one side of each electrode. As hydrogen enters on the anode side, the anode catalyst splits it into electrons and protons. The protons can pass through the electrolyte towards the cathode, but electrons can travel off and form an electric current to run an electric motor.

The catalyst at the cathode combines the protons migrating to the cathode through the electrolyte with the returning electrons through the motor and with oxygen from the air to form water. We can stack a number of such cells to provide higher voltages and more power.

German scientist, Gerhard Ertl, received the 2007 Nobel Prize for chemistry for his studies of processes on solid surfaces. His work has enhanced the process used to make fertilizer and the production of catalytic converters and hydrogen fuel cell technology. An interesting article in the October 2002 issue of *Scientific American* also discusses this technology.

Unlike an electric cell or battery, a fuel cell does not run down or require recharging. It operates as we continuously supply fuel and an oxidizer to the cell. A fuel cell power plant is up to 55 percent efficient, compared to a regular internal-combustion engine, which is only 30 percent efficient. It is a very clean technology, and the auto industry has even tested fuel cell cars.

Significant developments are in progress for commercialization of the technology. First, we need to develop better technology to produce hydrogen from water or other sources. Generation from the fuel gas causes pollution. We would need a hydrogen distribution system in addition to a production system. Other needed improvements for the commercialization of the fuel cell cars include lower costs, safer onboard hydrogen storage capacity, and increased fuel-cell life and durability.

KR Sridhar founded the Bloom Energy, which develops a flexible fuel cell system that produces clean, reliable, and affordable energy from a wide range of fuels. This technology enables consumers to generate their own electricity at a low cost and reduces their carbon emissions by 50-100 percent per kW, depending on the fuel. These Bloom Boxes are self-contained power-generating units designed for homes and businesses. Several companies including Google, Wal-Mart, Staples, and FedEx are testing them.

The Bloom Energy technology is based on fuel cells that are still expensive to produce. At the heart of Sridhar's device is a thin fuel cell made from a plentiful resource, sand. It is the size of a floppy disk, and it is painted with proprietary inks that allow the fuels to react with oxygen from the air, a chemical process that produces electricity.

Research to build compact rechargeable batteries that hold more power is also accelerating. Lithium and fluorine compound-based batteries are becoming common, because lithium gives us electrons and fluorine snatches electrons more easily.

The US Department of Energy has created the Applied Battery Research Program. They have been investigating several types of batteries for use in Electrical and Hybrid Electrical Vehicles, among them lithium-aluminum-iron-sulfide, nickel-metal hydride, lithium-ion, and lithium-polymer.

Lithium-ion batteries are the newest and fastest-growing rechargeable battery technology. These systems come closest to meeting all of the technical requirements, but they face four barriers for automotive use: short calendar life, low-temperature performance, low abuse tolerance, and high cost.

General Motors (GM) is coming up with a car that uses lithium batteries instead of a hydrogen fuel cell. This electric car runs on batteries and needs a charger to replenish the battery's power from an ordinary electrical outlet. A fully electric car can travel only 160 km. Car accessories such as air conditioning drain the battery even faster. The GM car is a hybrid in the sense that it can also run on gasoline after the battery power runs out. HEVs reduce the dependency on fossil fuels and improve the air quality because of fewer emissions. The government is encouraging their use through tax deductions.

Finally, researchers are also working on novel ideas such as making batteries from paper to power electronics in the future. The key to this new battery is the often-bothersome green algae known as cladophora. In fact, rotting heaps of this hair-like freshwater plant throughout the world can lead to unsightly, foul-smelling beaches.

However, this algae makes an unusual kind of cellulose, which has a very large surface area, one hundred times that of the cellulose found in paper. This allows researchers to increase the amount of conducting polymer available for use in the new device, enabling it to better recharge, hold, and discharge electricity.

- Wind Technology

We define wind energy as the "power generated by harnessing the wind." Uneven heating on the earth's surface causes wind. For example, the equator region receives more heat than Antarctica does. This creates a temperature gradient that causes wind to blow. Wind energy is the "force" of winds blowing across the earth's surface. The wind speed distribution over land is shown in Fig. 2.1.

The amount of kinetic energy due to wind within the earth's atmosphere equals about ten thousand trillion kilowatt-hours. According to research at Stanford University, wind power could satisfy the global energy demand seven times over—even if we can harvest only 20 percent of available wind energy.

Wind energy is free, clean, and non-polluting. The generation process does not produce any by-products that could be harmful to the environment. There are no chemicals involved and no waste production. Wind supply is plentiful, so wind energy is a renewable source. We can generate it continuously, during day and night. The only requirement is wind.

An efficient windmill can produce approximately 175 watts per square meter of propeller-blade area at a height of 25 m. In 2006, 73,904 MW were generated, so if each windmill has two square meters of area, that equals over 200,000 wind turbines working throughout the globe. Fig. 2.2 shows the amount of wind energy currently harvested by the entire world and a future prediction.

We associate wind energy with wind turbines and windmills. Long ago, we used wind windmills to power millstones, pumps, and forges. Even now, the British are constructing wind farms in upland areas of the British Isles, such as Wales and the Lake District. However, people are objecting to the construction because of visual and noise pollution.

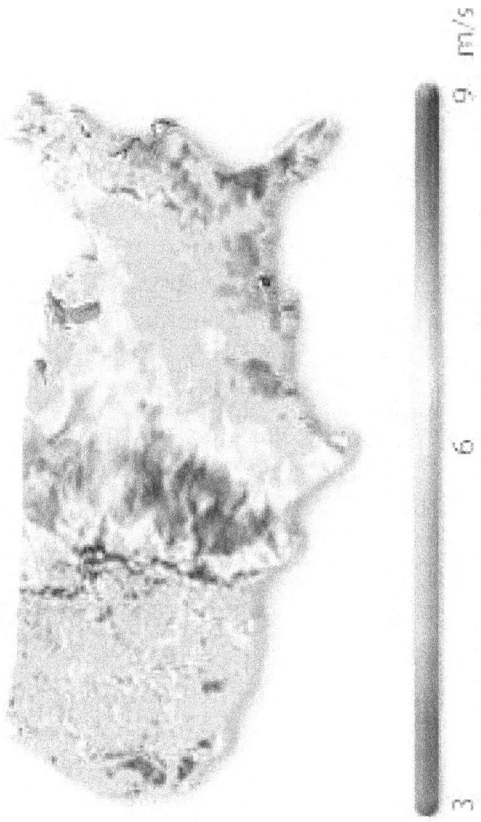

Fig. 2.1 - Wind Speed Distribution Over Land

To solve this problem, engineers have suggested moving to offshore locations away from the populated areas, where they would also get the benefit of higher wind speeds. An Ontario-based company is also working to an alternative approach by placing floating "turbines" hundreds of feet above the ground. The advantage would be a reduction in noise pollution, but it could pose a risk for airplanes or act as a giant conducting wire for lightning bolts.

Some of the advantages of wind technology are as follows:

1) It is renewable, non-polluting, free, and suitable for less sunny regions.

2) It can integrate well with other systems. We can use the generated wind energy full time in residential or commercial applications combined with our regular power supplies. It can also act as a backup in case our residential supply lines fail.

Some of the disadvantages are as follows:

1) Some people just do not like the look of windmills or wind turbines outside their window. A number of companies are working on solving this problem.

2) Wind farms generate noise in quiet, rural sites. Construction companies have tried to solve this by moving the turbines to unpopulated areas and offshore.

3) It is not very reliable due to the inconsistent nature of the wind. Construction companies try to

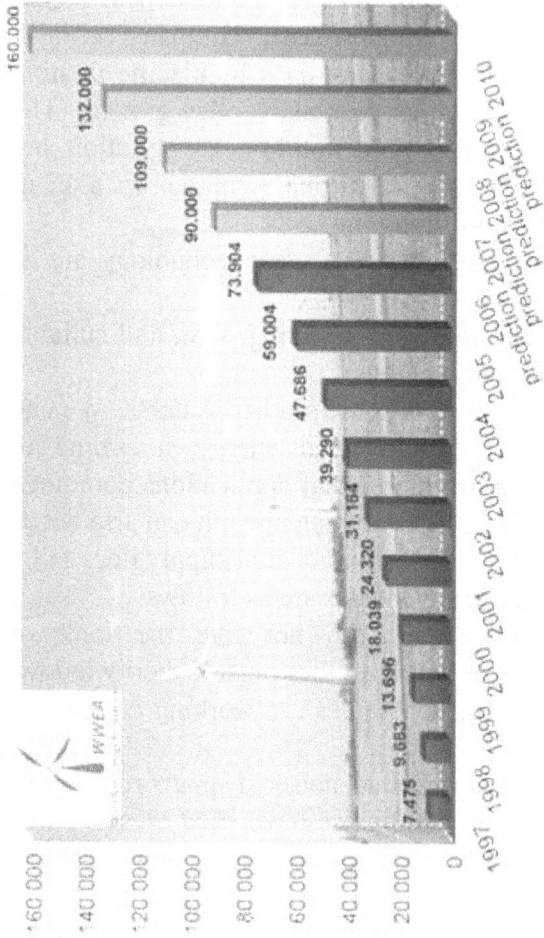

Fig. 2.2 – World Wind Energy
(US Department of Energy)

4) place turbines in the areas where wind blows most of the time.
5) It can affect the bird population. Birds and other flying creatures have trouble seeing the turbines. Special coloring patterns and slower moving blades could reduce this problem.

- Tidal Technology

Tidal/ocean energy comes from the regular rise and fall of the water level in the earth's oceans. Tides results from the gravitational forces between the earth, moon, and sun. They are big bulges of water created by the gravity of the sun and the moon. These bulges move around our planet, back and forth, creating currents and causing water levels to rise and drop.

From the position of moon, we can predict the rise or fall of water levels and their location. Usually, it takes about twelve hours for water to rise and twelve hours for it to fall. We can take advantage of this rise/fall phenomenon by harnessing renewable energy from it. By installing barrages (small dams), we can generate electricity using turbines.

Some of the advantages of tidal technology are as follows:
1) Because of this continuous rise/fall phenomenon, we can produce energy from tidal power.
2) It is clean, non-polluting, and truly renewable. We will not run out of it for two billion years when the earth's orbit gets too close to the sun and all of the oceans boil. It is reliable and

predictable. Unlike random weather patterns for wind or solar energy, the rise and fall of tides is more cyclic, which makes tidal energy more predictable.

3) Tidal turbines are up to 80 percent efficient in converting tidal energy to usable electricity. This is much higher than solar or wind energy generators. In our car engines, the conversion efficiency is only about 30 percent.

4) Barrages or small dams built for tidal energy could protect nearby cities or ship ports from dangerous tides during a storm. It would certainly reduce the damage.

Some of the disadvantages are as follows:

1) The initial investment to build a tidal energy plant is high, though the maintenance costs would be very low. Underwater tidal power is cheaper, since building dams is expensive.

2) Underwater tidal energy generation is still in the development stage, and we do not know the possible side effects to the environment due to placing turbines underwater. It is also difficult to predict the effect on ocean/river currents.

3) There would be a considerable effect on the ecosystem, since anything we place in the water can and does affect the ecosystem. The exchange of water volume between a basin and the sea can lead to the potential for increased pollution, as salinity of the basin decreases and sediment accumulation increases.

4) It would affect the fish population, even with the best barrage designs. The fish mortality rate per pass through the barrage would be about 15%. We have not yet found cost-effective solutions to this problem.

5) Finally, although tides are predictable, power stations only generate power when the tide is flowing in or out of the basin, which would only happen during certain times of the day. These systems can be placed virtually anywhere underwater, and they do not require special dams, channels, or underwater pathways to be constructed.

6) Systems appear to be safe for fish, as the rotors are slow turning. In addition, they should be safe for submarines, as submarines have complex radar navigation and do not easily run into things underwater.

- Biomass Technology

Biomass energy is the energy obtained from the methane gas generated by sewage or farm, industrial, and household organic waste, from specially cultivated organisms and trees. On a small scale, such plants are already functional in many countries around the world. However, use of the biomass technology on a large scale is yet to come. For example, growing algae on one hundred square miles can yield over forty million barrels of oil per day. This is twice the US daily consumption and about half the world's production.

Research activity in this field is heating up. There are about thirty-six companies working on producing energy from biomass. One can find lots of information on the web on sites such as www.algalbiomass.org. In addition, after producing coal-bed methane gas, a great deal of research is underway to use methogenes for energy production.

Methogenes are the tiny creatures that feed on minerals, carbon, and excreting methane from biomass. One could grow these creatures in sunlight and air (CO_2). Aquaflow Bionomics in New Zealand already operates a one hundred-acre form, and PetroSun has built an eleven hundred-acre algae farm in Texas.

We can also grow the algae in tanks, inject them into depleted coal-bed methane wells to feed these algae, and recover almost 50 percent of the energy left behind. As stated, there is potential for producing vast amounts of oil from the algae on farms in sunny places, with low-cost real estate.

We can improve the quality of algae to produce more oil from an algal culture through successive breeding. In fact, PerroAlgea Company is doing this, using an extremely greasy slime from the University of Arizona, developed with the DOE-funded Aquatic Species Program.

2.3 Technology & Modern Physics

WE now visit technology other than the energy conversion, based on modern physics. Let us first visit various semiconductor devices that form the basis of the modern revolution in technology. Next, we move on to the control of man-made processes and the environment. Then we visit the revolution in computer technology based on very large-scale integrated circuits (VLSI). We end our journey by visiting impact of modern physics on healthcare technology.

Developments in electronics and optics mostly come from quantum mechanics and the Standard Model. The progress in electronics and laser technology, based on quantum mechanics, has been truly remarkable. Developments over the past half-century in semiconductor devices, integrated circuits, compact and faster microchips, and laser technology have been particularly phenomenal.

Their applications in digital processing of signals and data inside a computer and error-free transmission have ushered in a revolution in communication and computer technology. It has made communication systems more compact and speedier. The advent of fast analog-to-digital converters and transmission of huge amounts of data at the speed of light have made digital technology practical. It has created new multibillion-dollar industries. Let us pay a brief visit to some of the significant developments that were based on quantum mechanics.

- Semiconductor Integrated Circuit Technology
The first significant device that ushered in the above revolution in technology was a transistor invented at Bell Lab in 1948. It was much smaller than a vacuum tube, it was silicon based, and it consumed less power.

A transistor circuit could do everything a vacuum tube circuit did. This was achieved by doping silicon with controlled impurities to create a shortage of valence electrons or "holes" (i.e., absence of electrons) that act as a sort of electron with positive charge and are ready Tto accept electrons.

For communication technology, a transistor circuit could transform and manipulate any signal in any manner. It could rectify (convert alternating current to direct current), amplify (boost the strength of the signal), modulate (mix message signal with carrier wave), oscillate (generate carrier wave signals), digitize (convert analog signal to digital bit streams), and so on. For computer technology, it could create two states, namely, current flow or no-flow, corresponding to binary digits one and zero.

The development of integrated circuit technology comes next. In this technology, we can etch complex circuits on a single silicon chip using a huge number of semiconductor devices such as transistors. For example, these days, we can put millions of transistors on an area of one square centimeter—the size of a fingernail.

For example, Intel plans to produce chips on 45-nanometer technology, which shrinks the circuitry's width to 45 billionths of a meter. The new family of chips will boast higher performance than previous generations partly because more transistors can be squeezed onto a single slice of silicon. Intel plans to introduce six different types of processors, including processors with four computing cores boasting eight hundred and twenty million total transistors. The development of such technology is very important for all modern applications since it reduced the size and the power consumption and increased the speed of operation tremendously.

- *Laser Technology*

A laser (light amplification by stimulated emission of radiation) is a device in which energized atoms release photons that have very specific wavelengths or frequencies corresponding to a color that depends on the excited state of the electrons. According to the atomic model, when an atom absorbs energy in the form of heat, light, or electricity, the electrons in its orbits may move from a lower-energy orbit to a higher-energy orbit. When these electrons return to the ground state, they release their energy as photons or light particles. This is the basic scientific principle behind light from the ordinary electric bulb, fluorescent lights, and lasers.

In lasers, we pump energy in a suitable medium with very intense flashes of light or electrical discharges, which creates a large collection of atoms in the excited state. In general, we excite the atom to a level that is two or three levels above the ground state. Two identical atoms with electrons in identical states release photons with identical wavelengths. Laser light is very different from normal light in the sense that laser light is monochromatic (single frequency). Laser light is also coherent (in phase). That is, each photon moves in step with the others. This means that all of the photons have wave fronts that move in unison.

The normal flashlight or light from an electric lamp is weak and diffused since it contains several frequencies and different waves are not in phase. A laser light is directional, very strong, and concentrated in a beam. Laser light contains one specific frequency or wavelength (one specific color). Theoretically, one can generate a laser field intense enough to break photons into electron-positron pairs.

The wavelength of light is dictated by the amount of energy that is released when an excited electron drops to a lower orbit. To obtain monochromatic coherent light from laser, we need stimulated photon emission, unlike the random emission in a flashlight. In stimulated emission, photon emission is organized.

It occurs when a photon encounters another atom that has an electron in the same excited state. The first photon can stimulate or induce atomic emission such that the subsequent emitted photon (from the second atom) vibrates with the same frequency and in the same direction as the incoming photon. Photons with a very specific wavelength and phase, reflected off a pair of mirrors at the ends of a suitable medium, travel back and forth in the medium.

In the process, they stimulate other electrons to make the downward energy flow and cause the emission of more photons of the same wavelength and phase. A cascade effect occurs, and soon, many photons of the same wavelength and phase come as laser light through the end where a "half-silvered," mirror lets some light through.

Depending on the medium—solid, gas, liquid, or semiconductor—we have different types of lasers.

a) Solid-state lasers have a solid medium (e.g., the ruby or neodymium [yttrium-aluminum garnet] "Yag" lasers). The neodymium-Yag laser emits infrared light at 1,064 nanometers ($nm\sim10^{-9}m$).

b) Gas lasers have a gas medium, and common gas lasers—helium and helium-neon, HeNe—emit visible red light. CO_2 lasers emit energy in the far infrared. They are used for cutting hard materials. Excimer lasers (the name is derived from the combination of terms excited and dimers) use reactive gases such as chlorine and fluorine mixed with inert gases such as argon, krypton, or xenon. When electrically stimulated, a pseudo molecule (dimer) is produced. The dimer laser produces light in the ultraviolet range.

c) Dye lasers use complex organic dyes, such as rhodamine 6G, in liquid solution or suspension as media. They are tunable over a broad range of wavelengths.

d) Semiconductor lasers, also called diode lasers, are electronic devices that are generally very small and use low power. They may be built into larger arrays, such as the writing source in some laser printers or CD players.

Laser technology has found many applications in the field of communication, computers, healthcare (e.g., surgery), and so on. Now, we can modulate, pack a laser-generated optical-wave with millions of signals, and transmit it at the speed of light, through a fiber optic cable instead of copper cables. To sum up, the developments in laser technology have been truly significant.

- Superconductor Technology

Many scientists believe that superconductor technology, based on the quantum phenomenon of superconductivity discussed in modern science, has the potential to usher in another industrial revolution. Superconductors offer no resistance to the flow of electricity.

Currently, we waste a lot of electrical power due to the resistance offered by the medium to the flow of electricity. It generates a substantial amount of heat in computers and electrical appliances. At present, we do have superconductors operating at extremely low temperatures.

If we could develop superconductors that operate at room temperature and avoid expensive refrigeration, it would be such a major advancement. We could develop new supercomputers by reducing heat generated in circuits. We could reduce power losses in transmission and in electrical appliances. We could make cheap power available for generating powerful magnets that could lead to the development of maglev trains, hover vehicles, new types of MRI machines, and more powerful particle accelerators.

The Japanese have already built a maglev train, the ML-500R. They tested it in 1971, achieving speeds of 517 km per hour on a 4.3-mile test track. The Germans have also developed such a prototype train, the TR-06. France's TGV conventional train, using a 25,000 horsepower engine, established a new speed record of 574.8 km per hour in 2007.

The high-energy particle accelerator electromagnets using superconductivity have already been used at the Fermilab Tevatron. Electronic devices such as SQUID (based on the Josephson Effect) have been used to measure the magnetic field of the body. It is capable of detecting even the weakest magnetic field variations.

Additional examples of technology based on the principles of quantum mechanics and atomic physics include health diagnostic tools such as X-rays, Computer-Aided-Tomography (CAT), Magnetic Resonance Imaging (MRI), Positron Emission Tomography (PET), and radiation therapy technology for the treatment of cancer. These will be discussed in the section on healthcare technology.

An interesting development was reported in 2007 in Nature Journal concerning another quantum phenomenon. In an experiment, a fleeting pulse of light was captured and then made to reappear in a different location by US physicists. This quantum phenomenon exploits the properties of super-cooled matter known as a Bose-Einstein condensate.

Though the emerging pulse was slightly weaker than the high-speed beam that entered the experimental setup, it was identical in all other respects. The work could lead to advances in computing and optical communication, according to Lene Vestergaard Hau of Harvard University and one of the authors of the Nature Journal paper.

- Nanotechnology

Nanotechnology has the potential to usher in, within a decade, a revolution in several areas of technology. These areas include consumer electronics through miniaturization, chemicals, and basic materials through the development of new materials, energy through thin film photoelectric and fuel cells, and pharmaceutical and medical technology through sensing, diagnostic, and therapeutic nanomedicine. Strictly speaking, it is the technology of the future, and we will visit its frontiers in the sixth chapter. However, it is so exciting that we briefly visit it here also.

What exactly is nanotechnology? Nanotechnology is based on nanoscience, which deals with the structuring of atoms and properties of materials at the nano-scale. Nanotechnology is about manipulating individual atoms, which is essentially a marriage of engineering and chemistry. The basic idea and goal of nanotechnology is very simple. From atoms, nature builds matter, and from matter, we build consumer goods. Nature can manipulate individual atoms and construct materials on a micro- $(\sim 10^{-6} m)$ and nano-scale-$(\sim 10^{-9} m)$. We are not able to do that.

Nanotechnology would provide us the capability to manipulate individual atoms and assemble them in a pattern to produce a desired structure. To get an idea as to how nature produces complex systems using techniques similar to nanotechnology, just look at the cells in our body that are built from certain atoms. These trillions of cells assemble our whole body according to a specified code in a DNA cell, and most of these cells are continuously replicated.

Nanotechnology tries to emulate nature. We already manufacture carbon nanotubes in research labs by rearranging carbon atoms in a hexagonal form. A carbon nanotube is extremely small (1/10,000th of the size of a human hair). It has the strength of a diamond, extreme flexibility (it can bend and twist, like steel), electrical and heat-conducting properties, and the ability to act as a semiconductor device. Because of such properties, this material itself is already finding many new applications. Carbon nanotubes have recently been used to kill cancer cells without harming healthy tissue.

In this technique, microscopic synthetic rods—carbon nanotubules—are inserted into cancer cells. The carbon nanotubes used by the Stanford team are only half the width of a DNA molecule. Thousands of such nanotubes can easily fit inside a typical cell. The surface of cancer cells is covered with receptors for a vitamin known as folate. The researchers taking advantage of this fact coated the nanotubes with folate molecules. It made it easy for the nanotubes to pass into cancer cells without binding with the healthy cells. The rods are exposed to near-infrared light from a laser, heating up and killing the cancer cell without harming the cells without rods.

Under normal circumstances, near-infrared light passes through the body harmlessly. However, the Stanford team found that if they placed a solution of carbon nanotubes under a near-infrared laser beam, the solution heated up to about 70° C in two minutes. Placing these heated tubes inside cells quickly destroyed them due to the heat generated by the laser beam. Exposure to the laser duly killed off the diseased cells, but left the healthy ones untouched.

Details of the Stanford University work are published by Proceedings of the National Academy of Sciences. As Hongjie Dai of the Stanford Research Team said, "One of the longstanding problems in medicine is how to cure cancer without harming normal body tissue. The standard chemotherapy destroys the normal cells along with cancer cells, and patients often lose their hair and suffer numerous other side effects. Intense research effort is directed towards finding a way to selectively kill cancer cells and not damage healthy ones."

Chad Mirkin, a researcher at Northwestern University in Evanston, Illinois, has developed technology to manipulate particles the size of atoms and molecules, which has led to novel inventions in medical testing and a high-resolution, molecule-based printing technique with applications ranging from electronics to drug discovery. Chad has been able to detect molecules in the bloodstream that are harbingers of disease with the help of gold nanospheres. This diagnostic test is about a thousand times more sensitive than any other system out there.

Geoffrey von Maltzahn at M.I.T has figured out ways to use tiny nanorods of gold to seek out and destroy cancer cells with none of the side effects (e.g., hair loss and nausea) associated with chemotherapies. Nanotechnology is poised to usher in a true revolution in many fields because of its unique features. We shall visit this topic again in the chapter on the frontiers of technology.

2.4 Control Technology

How did we change our environment and create the present industrial age? To reach the present stage of industrial development, we needed a very crucial technology, namely, control technology. Development of this technology was essential to control our environment, control the energy sources, control industrial plants, develop new materials, increase productivity, navigate, and send satellites into space.

Control systems are an integral part of almost all the physical and operational systems. In fact, all the functions of our body, even at the cellular level, are controlled by such systems, with the brain acting as the controller. Our brain is the controller for all our body functions and body systems. Our daily activities such as walking, talking, reaching for an object, and driving a car also use sophisticated control systems.

In fact, the concepts of control and feedback are critical for any useful application of science. Obviously, an uncontrolled application of technology cannot help us control our environment in a desirable manner and make us more comfortable. A control system is embedded implicitly not only in fabricated systems, but also in every phenomenon that we observe in nature, including human behavior.

We may not understand all the laws yet, but every phenomenon in nature is certainly controlled. Starting with the Big Bang, competing forces and physical laws control the expansion of the universe. Certain laws control and govern the cosmological constants, the formation of stars and galaxies, the formation of elements necessary for life, and our body systems, down to the individual cell levels.

Even when man tries to unleash uncontrolled processes, nature brings them to an equilibrium state and under control through adjustments of various parameters, which may not be to our liking. For example, nature always brings back to equilibrium the overpopulation through natural catastrophic adjustments. An uncontrolled process usually ends in a disaster. Humanity might not exercise proper control. Humanity might pollute the environment and disturb the ecological balance or wipe out all life on this planet using nuclear bombs.

For humanity, it would be a catastrophe, but planet Earth is just one planet. For the universe, which contains billions and billions of stars and most probably planets, it is not such a big deal. Violence in the universe is unbelievable. Just look at the galaxies colliding in the sky through a telescope.

The control systems used in environmental control, in the manufacturing industry, in airplane and satellite guidance, and in laser-guided and smart bombs can be quite complex.

The basic concept of an automatic control system is very simple. A block diagram of such an automatic control system is shown in Fig. 2.3.

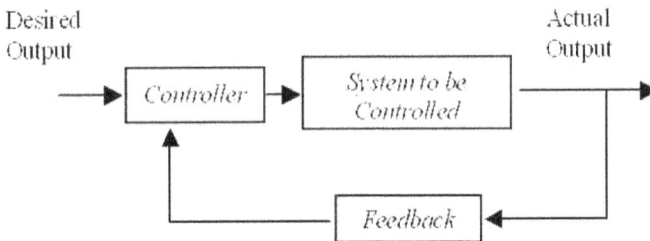

Fig. 2.3 – Basic Control System

The basic idea is the same whether we want to control the temperature, another environmental parameter, an industrial plant, a chemical process, an airplane or satellite guidance system, or an operational system such as an economic system. The objective is to obtain the desired outcome with the help of a controller by using the error signal, namely, the deviation of the actual output from the desired outcome.

The controller is the heart of a basic control system for controlling any process. The feedback loop continuously provides the information about the actual output from a process to the controller. The controller continuously compares the actual output of a system with the desired outcome. It keeps adjusting the input to the system with the help of controller until the desired outcome is achieved.

Earlier controllers were simple circuits that monitored the error signal (i.e., the difference between the desired and actual outcome). This error signal was used to change the input to the process to be controlled to minimize this error.

With rapid developments in control technology, the systems are now controlled with computers acting as controllers. We use such controllers to control simultaneously more than one parameter using a multivariable control system. Applications of such multivariable control systems are quite common in many industrial, chemical, and oil refining plants.

A control system can get complex in a hurry, as we demand more from it. We can always add more feedback loops to achieve more control. For example, we can control the system subject to variations in its parameters or due to external disturbances by providing additional feedback loops to monitor the variations in the state of the system.

We may not just want to control a process; we may want the "best" control, where the "best" is defined in terms of the importance given to a particular objective. Such an objective may be minimum cost, minimum time, or minimum energy expenditure to achieve the desired objective.

The optimality criterion also takes into account the existing constraints on the system. Alternatively, we can design controllers that can take action based on learning from the experience based on stored patterns. An example of such a learning control system would be using past experience when driving a car on icy roads. Conceptually, we can even think of designing a controller that would make intelligent decisions just as we do.

An interesting approach to designing the "best" control is based on the Bellman's Optimality Principle and dynamic programming. Simply put, it says, "Do the best you can from wherever you are, without worrying about the past." Its mathematics becomes quite involved, and one runs into the "curse" of dimensionality.

The number of options we need to analyze at every discrete step increases dramatically. Even the most powerful computers cannot handle such a vast number of choices. It is interesting to note, however, that our controller—the "brain"—uses a similar principle while making daily decisions.

Some progress has been made in the area of learning control systems and its applications in robotics. Neural networks that learn from past performance are used in designing sophisticated learning control systems. There is also some research effort in designing intelligent control systems for robots. However, before we can design intelligent systems, we must first understand the processes of thinking and decision-making that go on in our brains. Only then can we construct physical systems that will replicate this process.

Despite advances in expert and learning systems and intense activity in the field of artificial intelligence, we have a long way to go and we have to settle some very important technical and moral issues before we can produce intelligent systems.

Let us look back, and take stock of what we have visited thus far in the field of technology. Technology based on mechanical science and thermodynamics helped us amplify our physical power. Communication technology based on the science of electromagnetism extended our communication capabilities, as it enabled us to communicate over long distances.

Control technology helped us control nature and the power that we could harness from various resources. We now visit computer technology based on computer science and solid-state physics, which has ushered in an era where we are at the threshold of amplifying our intellectual capabilities.

2.5 Computer Technology

We have always been developing tools to improve our computing abilities. It started out with crude gadgets such as the abacus. Before World War II, we developed mechanical computers that used gears, levers, and cogs to perform arithmetic calculations. During World War II, we replaced mechanical devices with vacuum tubes.

The basic principle of computing with such devices is simple. Using binary bits (ones and zeroes) and binary logic, we can perform mathematical operations such as addition, subtraction, multiplication (an extension of the addition operation), and division (an extension of the subtraction operation). It appears that all things can compute. Almost any kind of material can serve as a computer. Human brains compute well; so do sticks, strings, and abacuses.

We can use for computation any device having two physical states corresponding to the binary bits. For example, we can convert the output from an electronic device to produce either a current flow or no flow in an electric circuit, corresponding to one or zero in binary logic, respectively. Thus, using a combination of such devices, we could perform calculations at a high speed since electrical signals travel fast. However, such computers required thousands of vacuum tubes, occupied large rooms, consumed a lot of power, and generated a lot of heat.

A block diagram, shown in Fig. 2.4, illustrates the basic components of a computer. The basic components and operations of a computer are simple to understand.

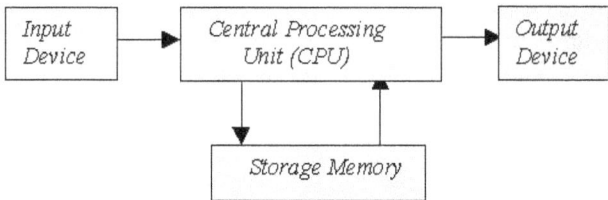

Fig. 2.4 – Basic components of a computer

1) The input device feeds the data and a series of binary-coded instructions (algorithm) to the Central Processing Unit (CPU), instructing the CPU to process the data in a sequential manner.
2) The CPU performs, according to the coded instructions, the arithmetic and other operations in the memory registers using binary logic.
3) The storage memory device stores data and supplies and receives information from the CPU in binary form since it can also have two states corresponding to the binary bits one and zero. Such a device could be electronic or magnetic (e.g., hard disk) or any other device that can have two distinct states.
4) The output device receives the output result of operations performed according to the instructions in the CPU. The output device either stores or prints the results.

Obviously, the performance of a computer, like any system, depends on its components. The speed at which a computer performs a computational task depends on several factors, such as:

a) how efficient the program or set of instructions is
b) how fast the input device can feed the instructions
c) how fast the CPU can process the instructions
d) how fast the memory devices can be accessed
e) how fast the output device can store or print the results

Earlier computers used paper tape and cards to punch data and instructions and crude input devices such as tape and card readers to feed instructions to the CPU. Nowadays, we often use input files that are already built either directly from various sensors in an experiment or on a keyboard to a personal computer (PC). A computer compiles and translates the input instructions, normally in higher-level computer languages, into a language that the CPU understands.

A computer's CPU only understands machine language consisting of sequences of binary bits, ones and zeroes. Nowadays, CPUs and memory devices have also become faster and more compact, thanks to developments in electronics and computer technology.

In 1948, the second revolution in computer technology commenced. With the invention of a silicon transistor at Bell Lab, the transistor replaced the vacuum tube. The third revolution in the computer industry was the development of the integrated circuit and microchip technology (large-scale integrated circuits on silicon chips).

We now have computers that can parallel process several sets of instructions simultaneously, computers that can communicate with each other, and computers and robots attached to sensors that can even appear to act, if not yet think, like us. The computers these days are extremely fast and can analyze many options.

French scientist Albert Fert and Peter Grunberg of Germany won the 2007 Nobel Prize for physics for discovering the phenomenon of "giant magneto-resistance" in which weak magnetic changes give rise to big differences in electrical resistance. This discovery has made it possible to miniaturize hard disks in recent years. The knowledge has allowed the industry to develop sensitive reading tools to pull data off hard drives in computers, iPods, iPads, and other digital devices.

Robert Dennard, an IBM engineer, invented a form of cheap, high-density memory that revolutionized the industry. DRAM, or dynamic random access memory, puts bits of data into capacitors. It is an energy storage device within a miniaturized electronic circuit, which periodically recharges the capacitors so that the information in them is not lost. The ability to put one bit on a single transistor was an improvement over previous six-transistor cells.

The invention of lasers and fiber optics technology has been another milestone in communication and computer technology. As stated earlier, photons can transmit, and even manipulate and store, information more efficiently than electrons can. However, all-optical computers with photon-based integrated circuits, quantum-based computing, and biological computers are not yet available in the market.

We discuss the status of these technologies in the section on the frontiers of technology. Lastly, since communication, computer, and control technologies are the lifelines of other technologies, recent developments in these technologies have deeply affected all other technologies.

2.6 Healthcare Technology

How can we live healthier and longer lives? We have always been on the lookout for technology that would help us in this regard. Since their first appearance, we have been fighting all kinds of diseases and treating them with all sorts of remedies. We are obsessed with prolonging our lives.

Some of the earliest systems of medicine are the "Ayurvedic" system that is described in ancient Hindu scriptures and the Greeks "Unani" system of medicine. The Ayurvedic system is related to Yoga and is based on intelligent and auspicious use of herbal medicine, diet, and daily practice. The Ayurvedic and Unani systems were essentially herbal-based. The chemical-based homeopathic system followed next.

However, the allopathic system is now practiced worldwide. In the modern allopathic system, medicines that treat various illnesses are usually developed from synthetic chemical compounds. These medicines are first developed in research laboratories, then tested on animals, and then tested on humans in controlled tests. After this process, the test data and results are submitted for approval to a government agency (e.g., in the USA, to the US Federal Drug Agency), which approves the drugs if they are considered safe. The pharmaceutical companies then make these approved drugs available to the public.

In the modern allopathic system of medicine, significant progress has also been made in identifying specific viruses or bacteria that cause certain illnesses. Based on this knowledge, vaccines have been developed that build antibodies in our immune systems that fight the undesirable guests attacking our bodies.

Such preventive measures keep us healthy. As we continue to learn more about the inner workings and chemical balance of our bodies, we can develop more medicines to maintain this balance and keep our bodies healthy.

Besides developing different types of vaccines and medicines to prevent or cure illnesses, medical technology has also evolved in a different direction. It concerns the technology for diagnosis and treatment of various diseases, which is based on atomic physics, electronics, and computers.

Examples of such technology are various diagnostic tools such as X-rays, Computer-Aided-Tomography (CAT) Scans, Ultrasound Scans, Magnetic Resonance Imaging (MRI), Positron Emission Tomography

(PET), and radiation therapy for the treatment of cancer.

A CAT scan provides different views of different cross sections of our bodies. It integrates multiple X-ray photographs taken through the body at different angles with the help of computers. Ultrasound scans use the fact that different types of tissues attenuate sound waves differently.

MRI uses the spin orientation of the nucleus of atoms in the presence of strong magnetic fields. When a high frequency signal is applied, it flips these nuclei upside down, and when reverted, it emits a small burst of energy. Different nuclei emit different signals, and various atoms in the body can be distinguished.

MRI uses the fact that the spin of the hydrogen nuclei can be 'flipped' from one state to another in the presence of suitable conditions. By measuring the location of these flips, we can get a picture of the location of hydrogen atoms (mainly as a part of water) in a body. Since tumors tend to have a different water concentration from the surrounding tissue, they stand out in such a picture.

Positron Emission Tomography (PET) is used for brain scans, and it can help diagnose various brain-related diseases such as Alzheimer's disease. Although these diagnostic tools have been very helpful, their resolution is still poor. For example, X-ray pictures are not sharp because X-ray beams are difficult to focus and manipulate. Current MRI devices cannot produce good pictures of fatty deposits in the heart because the deposits are thin and the heart is filled with fluids in motion. However, new technology is emerging fast in this area.

Another area in healthcare technology is based on the genetic revolution, where remarkable progress is being made. We visited the field of genetics in my book: Knowing the Unknown - I *Mysteries of Life - Past, Present, and Future.*

In 1944, Schrödinger wrote that life could be explained by a genetic code written in molecules within a cell. He was trying to explain life in terms of his quantum theory. Watson and Crick later proved his conjecture. They scattered X-rays off a DNA molecule and identified its double helical nature and atomic structure using X-ray crystallography. Since then, genetic science has evolved rapidly.

We now have the complete code of several living organisms, including insects, animals, viruses, bacteria, yeast, and so on. According to an agreement to be signed at the third International Barcode of Life Conference in Mexico City in 2009, hundreds of experts from fifty nations agree on a "DNA barcode" system that gives every plant on Earth a unique genetic fingerprint.

Today, we also have the complete map of the human genome—detailed information about the DNA molecule in a cell. The stage is now set for medicines based on gene modification and even for growing different body parts genetically.

As an example, let us visit the progress in today's research and treatment of cancer using various technologies that analyze and manipulate genes. Using DNA microarrays, researchers are attempting to study complex interactions of genes that are linked to cancer.

Other projects in genetics with important applications include the Genome Project data using bioinformatics, the study and cataloging of proteins, the study of human cells' built-in mechanisms (RNAi), and the creation of 3D models of proteins using X-rays.

Summary

During the start of the second phase of our journey, we visited important milestones in different areas of technology. Fig. 2.5 summarizes the development of technology, discussed in this chapter.

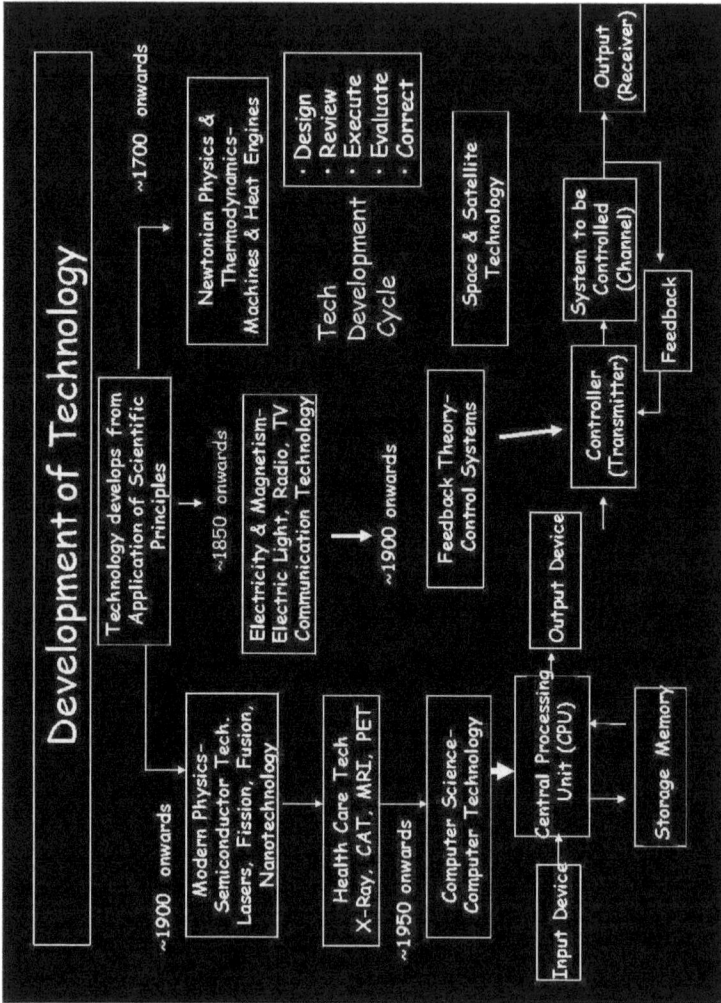

Fig. 2.5 – Development of Technology

Chapter 3
Top Ten Challenges

3. 1 Introduction

As we continue our journey, we discover that developments in technology parallel discoveries in science. Technology never stands still, since new technology comes along with developments in science. More developments are coming every day, and we see no end in sight. In fact, many more developments are bound to occur as we answer the remaining questions in science.

Since technology conceives what science perceives, most of the unknown technology is closely related to the unanswered questions in science. In this chapter, we first summarize the currently known technology. Then, we raise important questions and list the top ten challenges.

3. 2   Current Technology in a Nutshell

Let us summarize the currently available technology based on the well-known concepts in science.

1. Newtononian *mechanics*, based on Newton's laws, led to the development of powerful machines that could build other machines, which enabled us to amplify our mechanical efforts.

2. The development of the steam engine and other systems was based on the principles of

thermodynamics. This enabled us to use energy locked in various natural resources and provide faster transportation.

3. Scientific discoveries in *electricity and magnetism* led to the development of the electric bulb, electrical generators, and motors that ushered in the true industrial revolution.

4. Discoveries in the fields of electromagnetic waves, communication, and information science led to the development of communication technology that shows no signs of slowing down.

5. Developments in the Internet, satellite *communication*, signal processing, transmission, digital technology, and in related communication devices and systems technology have been truly remarkable.

6. Discoveries in *modern physics* led to energy conversion technology such as nuclear fission and fusion and hydrogen fuel cells. This technology promises to provide abundant and inexpensive energy for humanity.

7. Developments in *electronics*, semiconductor devices, integrated circuits, compact microchips with faster speeds, and laser technology have been phenomenal.

8. Developments in superconductor technology and *nanotechnology*, based on quantum mechanics, are continuing, which promise humankind reduced energy losses, powerful particle accelerators, new materials, and molecular machines.

9. The feedback theory led to the development of *control* technology. Control systems are an integral part of almost all physical, operational, and biological systems, down to the cellular level.

10. Development of this technology enabled us to exercise control over our environment, energy sources, and machines to improve quality, increase productivity, develop new materials, and navigate and send satellites into space.

11. Developments of *computer* science and component technology led to the development of computer technology. The development of the integrated circuit and microchip technology and the invention of lasers and fiber optics technology were very important milestones for communication and computer technology.

12. Developments in computer technology have affected many industries: telecommunication, printing and news media, the hardware industry, semiconductor and chip manufacturers, and the software industry.

13. Computers' marriage with communication and Internet has enabled us to interconnect millions of computers.

14. *Healthcare* technology and devices, based on discoveries in modern science, can monitor and diagnose health problems, treat various diseases, improve the quality of life, and prolong it.

15. Examples of such technology, based on atomic physics, electronics, and computers, include

various diagnostic tools such as X-rays, Computer-Aided Tomography (CAT) Scans, Magnetic Resonance Imaging (MRI), Positron Emission Tomography (PET), and radiation therapy technology for the treatment of cancer.

3. 3 More Questions & New Technology

In fact, so far, we have utilized only a few concepts in science that are essentially related to transforming energy and materials from one form to another. We have yet to conquer all the forces of nature. Despite rapid developments in technology, we are far from gaining complete control over our environment.

We may never succeed in completely controlling the weather or huge forces of nature such as hurricanes and tornadoes. We also cannot yet control gravity. In communication, we still cannot effectively communicate a 3D video.

In computing, we do not have "smart" machines that can think like us. In the field of energy, we are still using fossil fuels for transportation. In medical technology, we cannot prevent or treat all forms of cancer or many other diseases. Many expectations remain unfulfilled, and several questions remain unanswered.

As we continue our journey, more questions arise concerning the present state of technology. Specifically, we raise unanswered questions in technology based on Newton mechanics, thermodynamics, communications, modern physics, quantum mechanics, control, computers, and genetics. The search for answers to these questions is bound to lead to the development of technology that is unknown at present.

- Classical Mechanics and Thermodynamics

We have noticed that classical Newtonian mechanics and the laws of motion deal with the interplay of matter and forces. They have enabled us to build powerful machines, which can amplify our physical efforts through the application of mechanical energy transformed from energy locked in various natural resources.

Thermodynamics is concerned with thermal or heat dynamics. An understanding of heat energy led to the development of heat engines. It also led to the technology for converting other forms of energy into mechanical energy and brought in new factories and an industrial revolution.

The technology based on this aspect of science has reached maturity. However, new and improved unknown technology remains to be discovered. For example, one hears about the development of new engines and modes of transportation with faster speeds and lower energy consumption.

In this regard, the following questions concerning the development of technology remain unanswered:

1. How far can we go to develop more efficient energy conversion and energy storage techniques?

2. How far can we go to improve the energy utilization efficiency by controlling the factors that waste energy?

We also need to answer questions concerning gravity and the law of gravitation raised in the second chapter on science. Then, we might be able to answer the following questions:

1. Can we control and use gravitational fields more effectively?
2. Can we develop anti-gravity machines?

- Electricity, Magnetism, and Light Waves

In the lighting industry, the market is flooded with new types of long-life electric bulbs and other lighting devices. We are also witnessing a revolution in communication technology. The telephone industry is undergoing a revolution. We already communicate at the speed of light through electromagnetic waves and send audio/video signals between any two points on this planet, between Earth and objects far out in space, and under the oceans. The electromagnetic spectrum is getting quite crowded.

Nevertheless, we are still searching for the answers to the following questions:

1. Are there any limits to communication in terms of speed, distance, and type of signals?
2. Have we exploited all possibilities and discovered all means of communication?

With regard to the first question, communication is currently limited by the speed of light, signal attenuation, and distortion in the system, including channel noise and available bandwidth. We have developed some clever schemes using digital technology to overcome the signal-to-noise limitations. However, light speed's limit restricts the speed of communication. For example, it would take billions of years to receive information about what is happening today in the farthest corners of the present universe.

Alternatively, what we observe there today happened not today, but billions of light years ago—very old information, indeed! We still do not have the technology to communicate three-dimensional dynamic video signals.

To see stereoscopic 3D, one must send a right-eye view to the right eye and a left-eye view to the left eye. It creates the 3D effect in the brain once the person's eyes see these two views. Hyundai in Japan and LG Electronics in South Korea recently announced polarization technology, called X-pol (for "cross-polarization"), for viewing 3D LCD high-definition TVs.

The polarized glasses separate images for the right and left eyes from the even and odd horizontal lines of video. This technology might become available soon. Samsung and Panasonic did announce that they are introducing 3D television sets in 2010, but this technology has a long way to go.

Regarding the second question, some people talk about extrasensory perception (ESP), telepathy, teleportation (transmitting a solid object atom-by-atom over space), and other means and channels of communication. We are bound to develop yet unknown new technology as we probe deeper into the unanswered questions of science.

- Energy Conversion & Modern Physics

We have already described some of the remarkable developments in energy conversion and other technology based on modern physics. For example, we discussed fission and fusion technology, quantum mechanics, and technology based on the Standard Model such as semi-conductor and integrated circuit technology and laser and superconductor technology.

The unanswered questions in this type of technology parallel the unanswered questions in modern physics listed in the previous chapter. Some related and additional important questions that will deeply affect the development of yet-unknown technology are as follows.

1) Is it possible to develop technology for the safe disposal of nuclear waste from nuclear fission?
2) Is it possible to develop fusion technology for power generation?
3) What more can we do with solar, wind, and tidal wave energy?
4) Is it possible to develop devices for quantum computers?
5) Can we develop superconductor materials that can operate at room temperature?
6) Can we build special devices for nano assemblers, replicators, and other nano machines?

The future of technology and, in some sense, our own future rests on the answers to these questions. All these questions most likely will be answered in the affirmative. The question is not if, but when. When these questions are answered in the affirmative and the development of new technology starts, it will truly usher in another industrial revolution, which is even hard to fathom now.

- Control Systems & Robotics

We have seen that control systems are at the heart of almost every industrial application. A process has to be controlled at some stage if it is to help us and be of some use. We also talked about the adaptive, learning, and intelligent control systems. Such control systems adapt to changes in the plants/processes we want to control, store action patterns based on learning experience, utilize such experience, and take "intelligent" control actions on their own.

Furthermore, as we move forward in the area of intelligent control systems and robotics, we have to be careful. What values or criteria should we assign for the optimal design of robots? How do we ensure that the fabricated intelligent systems or robots will not turn against humanity and harm us?

Since robotics is a fast developing area of technology, let us examine it a bit more closely. Isaac Asimov, a distinguished science fiction writer, gave three laws of robotics to encode robots so they would not harm their human masters. These laws state the following:

1) A robot may not allow injury or harm to a human being through its action or inaction.
2) It must obey the orders unless they conflict with the first law.
3) A robot must protect its own existence provided it does not conflict with the first and second laws.

One obvious problem with such laws for a robot or computer is defining the terms "injury" or "harm." Complex pattern recognition and common sense (not so common, even in humans) are two of the primary requirements for intelligent thinking.

Present-day computers or controllers do not yet have such capabilities. They do not have what we call a "conscience" or "self-awareness." The most important question we should ask is whether computers could have such capabilities.

To sum up, the questions regarding control system technology are as follows:
1) In control system technology, how far can we go?
2) Can we design intelligent control systems?
3) In robotics, how far can we go?
4) Can we design automatic biological control systems similar to those that nature provides for living entities?

- Computer Systems

As mentioned in the previous chapter, computers started out as simple mathematical toys, but developments in computer technology have been rapid and overwhelming. This has affected many industries such as telecommunication, printing, news media, the hardware industry, semiconductor, and chip manufacturers, and the software industry. Today, we find computers everywhere—in factories, offices, and homes. The impact of computers in our everyday lives is hard to fathom.

Computers' marriage with communication and Internet technology has enabled us to connect millions of computers. It has brought the world closer than we can imagine. We are at the verge of a technological revolution that shows no signs of slowing and knows no bounds.

We make some predictions about its future in the next chapter, but here, we pose the following questions that remain unanswered.

1) What are the size and speed limitations for computers?
2) Can we build computers using nanotechnology or quantum chips?
3) Can we develop an "intelligent" or "smart" computer that can think like us?
4) Can we build advanced biological computers that are similar to the human brain?
5) Can we design intelligent computers that can amplify our intelligence (similar to the mechanical machines that amplify our mechanical capabilities)?

- Healthcare

We have already discussed the remarkable advances in this technology. We now can monitor and diagnose problems, treat various diseases, improve the quality of life, and prolong it. We have already discussed in some detail the revolution in genetics and molecular biology. For now, confining our discussion to healthcare technology, we raise the following questions:

1) In healthcare, how far can we go in developing new technology to monitor and improve our health?

2) Can we develop healthcare diagnostic devices that give an exact 3D dynamic picture of inside the body?

3) Can we cure cancer and other diseases using genetics?

4) How would the merger of biology and nanotechnology affect healthcare?

5) More specifically, what would be the impact and implications of the gradual inclusion of molecular machines based on nanotechnology in the body?

3. 4   Top Ten Challenges for Technology

We present here a list of the top ten challenges for technology.

1) In energy, we have developed several forms of energy conversion, but can we get away from dependence on the limited supply of fossil fuels and develop new and more efficient forms of renewable energy conversion techniques?

2) In nuclear energy, we generate power from fission technology, but can we develop technology for the

safe disposal of nuclear waste or develop cost-effective nuclear fusion technology?

3) In matter, we have developed many new materials: steel, plastics, polyesters, composite and nano-carbon material, and so on. However, can we develop technology to build new materials with any properties that we desire?

4) In device technology, we have developed many devices: semi-conductor and integrated circuits, lasers, and superconductors. However, can we develop quantum devices for quantum communication and computers and develop superconductors that function at room temperature?

5) In communication technology, we can communicate signals via satellites and across space, limited only by the speed of light. Can we break this speed barrier, and develop better means of recording and communicating 3D pictures/video?

6) In transportation technology, we can travel vast distances on land, sea, and in air. However, our range and speed is limited. Can we remove some of these barriers by developing new types of fuels and vehicles?

7) In control technology, we have developed control systems for industrial processes and spacecraft. However, can we develop learning and intelligent control systems and robots that can exercise "intelligent" control on their own like us?

8) In computer technology, developments in computer technology have affected every industry and our daily lives. However, can we develop "intelligent" or "smart" computers that think like us?

9) In healthcare technology, we have made remarkable advances in technology and can monitor and diagnose problems, but can we develop healthcare monitoring and diagnostic devices that produce 3D dynamic pictures of inside of our bodies?

10) In the medical field, we have been able to treat various diseases. However, can we develop effective gene-based technologies to prolong life, cure cancer, treat compulsive behavior, and grow different organs of the body?

Fig. 3.1 lists the unsolved problems and the unknown technology that awaits us.

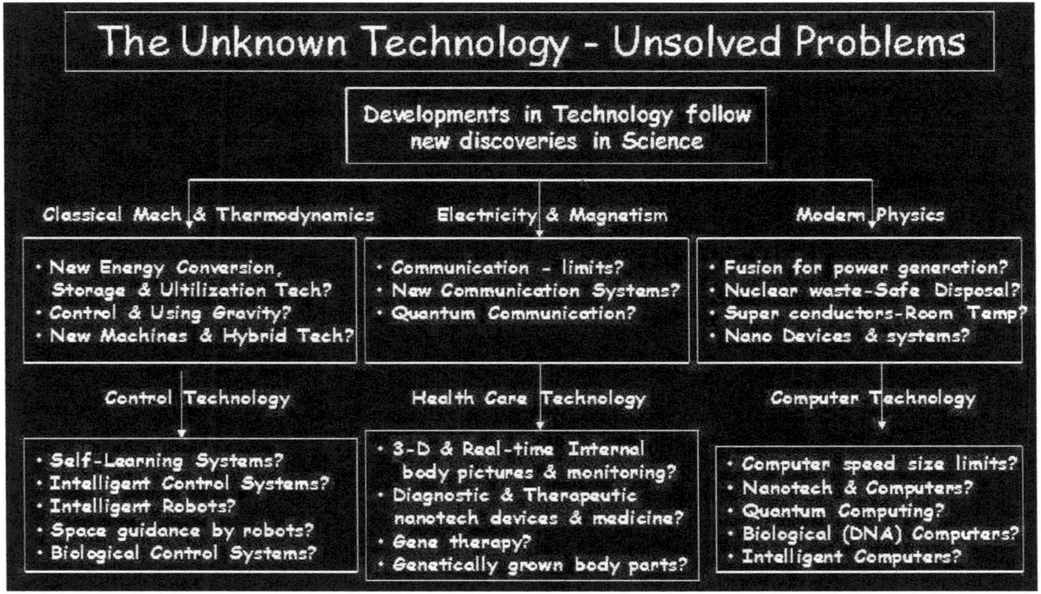

Fig.3.1 – Unsolved Problems & New Technology

The Unknown Technology - Unsolved Problems

Developments in Technology follow new discoveries in Science

Classical Mach.& Thermodynamics
- New Energy Conversion, Storage & Utilization Tech?
- Control & Using Gravity?
- New Machines & Hybrid Tech?

Control Technology
- Self-Learning Systems?
- Intelligent Control Systems?
- Intelligent Robots?
- Space guidance by robots?
- Biological Control Systems?

Electricity & Magnetism
- Communication – limits?
- New Communication Systems?
- Quantum Communication?

Health Care Technology
- 3-D & Real-time Internal body pictures & monitoring?
- Diagnostic & Therapeutic nanotech devices & medicine?
- Gene therapy?
- Genetically grown body parts?

Modern Physics
- Fusion for power generation?
- Nuclear waste-Safe Disposal?
- Super conductors-Room Temp?
- Nano Devices & systems?

Computer Technology
- Computer speed size limits?
- Nanotech & Computers?
- Quantum Computing?
- Biological (DNA) Computers?
- Intelligent Computers?

Chapter 4
Frontiers of Technology

4.1 Introduction

As we walked through the evolution of technology and raised several questions, we must have realized that we can still go a long way. In this chapter, we visit frontiers of communication technology, quantum communication, computing, nanotechnology, and genetics and review the efforts to develop intelligent control systems, robots, and computers based on the study of brain processes.

As we review the progress, we observe that we are bound to develop many new technologies. The unanswered questions in each technology define the current frontiers of that technology. The current activity on the frontiers will decide the future of technology. Fig. 4.1 summarizes our search for new technology.

We visit and focus here on frontiers that can drastically alter the future of technology and bring about revolutionary change in our lives. We visit first specific areas on the frontier of communication technology. The activity on this frontier has far-reaching implications.

After visiting the frontiers of teleportation and nanotechnology, we visit the frontiers of computer technology. From there, we move on and visit the frontiers of control technology, especially as it relates to simulating human intelligence. During this part of the journey, we also visit several frontiers of technology based on modern physics. Finally, we visit the frontier of healthcare technology, with specific reference to revolutions in molecular biology and genetics.

4.2 Frontiers of Communication

We have already developed various communication systems that can transmit information from one spot to another. The US Army is even developing a technology known as synthetic telepathy that would allow someone to create email or voicemail and send it by thought alone. One essentially reads the electrical activity in the brain using an electroencephalograph, or EEG.

Information is an essential element of the universe. As stated earlier, some scientists claim that even hard matter is information. They claim that particles like electrons, ions, and atoms have certain properties, and if we can transmit this information and reproduce the properties of quantum particles making up an object in another particle group, we could precisely duplicate the object. We need to transmit only the information about the particles' properties and not the particles themselves.

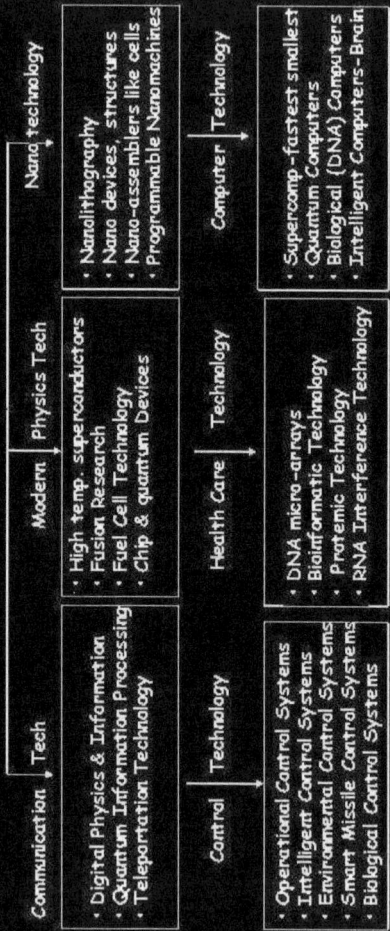

Fig. 4.1 – Frontiers of Technology

Search for New Technology - Frontiers

Unsolved Problems define
Frontiers of Technology

Communication Tech
- Digital Physics & Information
- Quantum Information Processing
- Teleportation Technology

Modern Physics Tech
- High temp. superconductors
- Fusion Research
- Fuel Cell Technology
- Chip & quantum Devices

Nano technology
- Nanolithography
- Nano devices, structures
- Nano-assemblers like cells
- Programmable Nanomachines

Control Technology
- Operational Control Systems
- Intelligent Control Systems
- Environmental Control Systems
- Smart Missile Control Systems
- Biological Control Systems

Health Care Technology
- DNA micro-arrays
- Bioinformatic Technology
- Proteomic Technology
- RNA Interference Technology

Computer Technology
- Supercomp-fastest smallest
- Quantum Computers
- Biological (DNA) Computers
- Intelligent Computers-Brain

- *Digital Physics & Information*

According to Wheeler, who coined the term "black hole," atoms are essentially made up of bits of information. In a 1989 lecture, Wheeler said, "Its are from bits...every particle, every field of force, even the space-time continuum itself — derives its function, its meaning, its very existence entirely from binary choices, bits. What we call reality arises in the last analysis from the posing of yes/no questions."

For example, to form a water molecule, picture two atoms of hydrogen and one of oxygen. As they come together, each seems to be calculating the optimal angle and distance at which to attach to the others. The oxygen atom uses yes/no decisions to evaluate all possible courses toward the hydrogen atom. Then, it selects the optimal 104.45° (most probably according to the optimal principle of universal change, discussed in the first two volumes of the series: Knowing *the Unknown - I & - II*, and moves toward the other hydrogen at that very angle. Every chemical bond is thus calculated.

According to Kelly's "God is the Machine" article in the December 2002 issue of *Wired* Magazine, if the theory of digital physics holds up, movement ($F = ma$), energy ($E = mc^2$), gravity, dark matter, and antimatter can all be explained by elaborate programs of one/zero decisions. Bits can be seen as a digital version of the Greek "atoms"—the tiniest constituents of existence. These new digital atoms are the basis, not only of matter, but also of energy, motion, mind, and life. Kelly says that in a world made up of bits, physics is the same as a simulation of physics.

Some biologists have proposed that life itself might be information. The entire human genome sequence (three gigabytes) represents the prime coding information of a human body (i.e., our life as numbers). It implies that biology can be conceived by science as an information process, and life itself is information. We can put the entire three billion digits of our DNA on about four CDs at today's rates of compression. If computers keep shrinking, we can imagine packing information numerically about our complex bodies to the size of two micro-memory devices, which could be called the egg and sperm.

However, starting with the union of a male sperm and a female egg, nature takes about nine months to make a baby in a mother's womb. It will remain beyond our capabilities for a long time to simulate such a complex process, since we do not even have a complete understanding of these processes that take place inside a womb. As discussed later, disassembling a solid object and reassembling it at another place atom by atom will also remain beyond our capabilities for the near future.

Scientists have an amazing ability to observe apparently diverse phenomena and then formulate a single concept, law, or equation that unifies them all. The simplest example would be Newton observing an apple falling to the ground or the moon orbiting the earth and coming up with the law of gravitation that could satisfactorily account for all such phenomena. Another such example would be the development of the laws of thermodynamics governing energy. In the field of information science, Shannon, in 1948, unified all phenomena related to classical information processing by quantifying the information content produced by an information source.

Shannon defined it to be the minimum number of bits needed to store reliably the output of the source. As stated in the first part of this book, his expression, known as Shannon entropy, plays a central role in data compression, information transmission over noisy channels, and the analysis of stock market behavior or the behavior of any system that processes information.

The reason for revisiting Shannon's work on information is obvious. According to the preceding discussion, the most important concept in this universe appears to be general information processing, namely, coding, transmission, and information reproduction. As we learn more about the true nature of information, we can develop technology that can process information more efficiently and at a faster pace.

We are now at the verge of developing quantum information science that will unify the concept of information at quantum level. Such a theory would be analogous to the theory of energy in thermodynamics and classical information theory. It would be applicable to such diverse phenomena as teleportation, quantum entanglement, quantum cryptography, superconductivity, and so on.

- Quantum Information

Let us visit the frontier of quantum information science, which could directly lead us to quantum computers. Nielsen's article "Rules for a Complex Quantum World" in the November 2002 issue of *Scientific American* gives a fascinating account of developments in this area.

As we know, the basic source of classical information is the bits, zero and one, abstracted from the principles of classical physics. Quantum information science generalizes bits to quantum bits, or qubits, abstracted from the principles of quantum mechanics.

Table 4.1 compares classical information science to the fast-emerging quantum information science.

Information	Classical Information	Quantum Information
Basis	Classical Physics	Quantum Physics
State	Classical State **bit** 0 or 1	Quantum State **qubit** - 0, 1 & superposition of 0 & 1
Physical Representation	Magnetic regions, electric signals, etc.	Spin of atomic nucleus, polarization of light
Important Concept	Shannon's Entropy	Qubits Entanglement
Problems	Limited information capabilities– only two states	Preserving fragile quantum state, qubits inaccessible hidden information
Applications	Information transmission, compression, processing, computers	Super condense coding, cryptography, quantum computers, teleportation,

Table 4.1 Classical vs. Quantum Information Science

In quantum mechanics, an object can not only have one of the two different states, but also additional states. A qubit can thus have many more possible states than just a bit, zero or one. These qubits can exist in superposition and simultaneously involving both zero and one. For example, if a classical single bit zero represents north, and one, the South Pole, then a qubit state corresponds to points on a surface of a sphere whose longitude and latitude information is expressed as a superposition of binary bits.

We can encode a qubit with an infinite amount of classical information through arbitrary precision. Some of the important problems in quantum information science and their resolutions are presented as follows.

- Understanding Quantum vs. Classical World

Quantum mechanics predicts that an object can exist in two states simultaneously. However, such states, in which an object is effectively in two places at once, have only been accomplished with single particles, atoms, and molecules. In 2010, in an experiment reported in the journal *Nature*, scientists produced a quantum state in an object billions of times larger than previous tests. This experiment could have significant implications in quantum computing.

According to quantum mechanics, objects absorb and emit energy in tiny discrete packets known as quanta. If one removes all the energy that an atom gets from the jostling atoms in its environment by cooling it to phenomenally low temperatures, it can reach its "quantum ground state," and no more energy can be removed. If one then returns just one quantum of energy, one can say that the atom is in two states at the same time. Although only one quantum of energy is put in, any measurements will show either one or zero quanta. Thus, the atom is said to be in a superposition of states.

In the experiment reported in *Nature*, Professor Cleland and his team used an object just big enough to be seen with the naked eye. They used a tiny piece a piezoelectric material, which expands and contracts when an electrical current is run through it.

Researchers first brought the material to the quantum ground state through super cooling. Then they "pumped in" just one quantum of electrical energy at a time. They observed the oscillator begin to vibrate as it converted that added quantum into one quantum of vibration energy. As it vibrated, the team showed that the resonator was in one of those superpositions of states, with both one and zero quanta of energy.

- Extracting Hidden Information from Qubits

The information in a qubit must be extracted by a measurement. The principles of quantum mechanics tell us that we can observe or extract only a single bit of information from a qubit, no matter how we encode a bit or measure it. Halevo proved this result in 1973.

Treating hidden information as a unit of quantum information, he developed a mathematical concept, known as the Halevo chi, which provided simplification similar to Shannon's entropy to simplify the analysis of more complex phenomena. This concept has been used for compressing quantum states in a manner analogous to classical data compression.

- Understanding Quantum Entanglement

A group of two or more quantum objects can have entangled states. Entangled objects behave as though they are connected through space, and distance does not attenuate entanglement. A measurement of an entangled object simultaneously provides information about the entangled partner. Quantum entanglement refers to the way observations of two particles correlate after interaction. The entanglement takes place because the conservation laws are exact but most observations are probabilistic.

Entanglement properties are independent of physical representation (i.e., an entangled photon pair, entangled atomic nuclei pair, or even a photon entangled with a nucleus can perform the same task).

A general theory is not yet developed for the quantum entanglement concerning information, similar to the laws of thermodynamics applicable to energy regardless of the form of energy. Such a theory, when developed, might also shed some light on the behavior of condensed matter such as, high temperature superconductivity that involves quantum entanglement.

A major problem in "entanglement" experiments has been the inability to pass this information reliably. Jian-Wei Pan and colleagues at the University of Vienna showed that we could confirm the occurrence of teleportation by monitoring the outcome of the interaction that teleports the qubit. If both photons are detected as expected, it is safe to assume that the quantum state has been teleported.

However, in practical experiments, the apparatus involved was not sensitive enough to prevent stray photons from causing frequent "false positives." To lower the number of photons in the system and thus the false positives, they reduced the intensity of the source used to fire photons at the entangled pair.

According to them, if the two interacting photons are detected in this set up, it is more than 97 percent certain that the state has been teleported. This accuracy is enough to extend the use of quantum communication by the use of "quantum repeaters," which can teleport qubits over long distances.

Although the quantum entanglement phenomenon is not completely understood, we can still make use of it. Ekert at Cambridge showed that we could use it to distribute cryptographic keys, preventing eavesdropping.

In 1992, Bennett and Wiesner showed that we could utilize it to send classical information from one place to another, using the super condensed coding process. We also have had some success in developing a measure and a standard unit of entanglement.

In 1995, Shor and Stean at Oxford were also able to develop an error-correcting scheme for quantum information, which is similar to those used for eliminating transmission errors in classical information. In short, we are all set on the path leading to quantum communication.

4.3 Frontiers of Teleportation Technology

In transportation, normally, the objects are physically transported from one place to another. New technology is continually developed in this area of transportation. Japan is making great strides in developing cars that run on hydrogen fuel cells and trains that levitate on magnets using superconductors.

Some scientists are even boldly exploring a frontier that is currently in the realm of science fiction. Many of us have watched *Star Trek*, where the crewmembers of the Starship Enterprise step into a transporter room, vaporize, and then reassemble elsewhere.

However, Hawking recently discounted science-fiction ideas from programs such as Star Trek, like using warp drive to travel at the speed of light, for taking humanity to a new outpost. Instead, he favored "matter/anti-matter annihilation" as a means of propulsion. When matter and anti-matter meet, they disappear in a burst of radiation. If this was beamed out of the back of a spaceship, it could drive it forward.

Some scientists are asking the ultimate question: is such teleportation within the realm of possibility? IBM lab in the United States provided the theoretical basis in the area of teleportation in 1993, and several physicists in quantum mechanics became interested in this area.

About forty laboratories worldwide have been experimenting in this area. In June 2002, a team of physicists at the Australian National University (ANU) announced that they had successfully disembodied a laser beam in one location. Then they rebuilt it in a different spot, about one-meter away, in the blink of an eye. The laser beam was destroyed during teleporting using the process of quantum entanglement.

After the successful experiment at ANU in Australia, the project, leader, Lam told a news conference that there was a close resemblance between his team's achievement and the movement of people in the science fiction series *Star Trek*. However, the reality of beaming human beings between locations was still light years off since humans, made up of zillions of atoms, posed a near-impossible task.

"In theory, there is nothing stopping us from doing it, but the complexity of the problem is so huge that no one is thinking seriously about it at the moment." However, Lam said science was not too far from being able to teleport solid matter from one location to another. According to him, their breakthrough opens enormous possibilities for future super-fast and super-secure communications systems, such as quantum computers over the next decade.

4.4 Frontiers of Nanotechnology

In a previous chapter, we visited developments in several technologies based on modern physics. The frontiers of all such technologies—namely, nuclear fission, fusion, hydrogen and solid fuel cells, semiconductors, lasers, superconductors, and nanotechnology—are being extended continuously. For example, the "brain-machine interface" developed by Hitachi Inc. analyzes slight changes in the brain's blood flow and translates brain motion into electric signals. Such a signal was used to move a toy train set via a control computer and motor during a recent demonstration at Hitachi's Advanced Research Laboratory.

Underlying Hitachi's brain-machine interface is a technology called optical topography, which sends a small amount of infrared light through the brain's surface to map out changes in blood flow. We are bound to have some startling developments in several areas in the near future.

However, we explore and focus here only on the frontier of nanotechnology, which is truly revolutionary. As discussed in the evolution of technology, we are about to enter a new and exciting era of nanotechnology. Nanotechnology holds the potential to affect all the other technologies in the near future in ways that we cannot even imagine.

We could have supercomputers that we can carry in our pockets. In fact, using nanotechnology, we could manufacture machines with unbelievable capabilities that are the size of molecules. Some of the potential applications of nanotechnology are as follows:

1) Potential revolution within a decade in the areas of consumer electronics through miniaturization, chemicals and basic materials through new materials, energy through thin film photoelectric and fuel cells, and pharmaceutical and medical technology

through sensing, diagnostic and therapeutic nanomedicines.

2) Development of nanorobots programmed to perform delicate surgeries, attack and modify the molecular structure of cancer cells and viruses to make them harmless, and slow down or reverse the aging process.

3) Development of nanomachines to bring about positive environmental changes, removing contaminants from water sources, cleaning up oil spills, manufacturing materials with less pollution, and reducing our dependence on non-renewable resources.

4) Development of molecular computers, new generations of computer components, and storage devices that can store trillions of bytes of information in a structure the size of a sugar cube. High-density memory projects underway by several companies are expected to bring nanodrives in the market starting in 2005.

Suppose we wanted to build a macro-structure following the lead from nature. We must be able to manipulate individual atoms.

IBM scientist showed in 1990 that it is possible by positioning 35 xenon atoms on the surface of a nickel crystal using an atomic force microscopy instrument and spelling the word IBM. The Dip Pen Nanolithography (DPN) technology developed by NanoInk Inc. can enable us to build nanoscale structures and patterns by literally drawing molecules onto a substrate.

We need more such tools and infrastructure—picks and shovels—to do nanotechnology. What would it take to make nanotechnologya reality? The following things must happen.

1) We must start with passive materials and then move on to structures and devices and then to nanosystems. Like any technology, we cannot force a new technology application unless we demonstrate the value of nanotechnology in terms of its effectiveness, efficiency, or cost savings.

2) To build large-scale systems, we must be able to build trillions of assemblers or nanoscopic machines like cells in our body, which can be programmed to manipulate individual atoms.

3) We must be able to build replicators or nanomachines, which can be programmed to build more assemblers, just as the cells are replicated in the body.

The first two developments could happen very soon. The remaining developments are far away, although some scientists at NASA think otherwise. We might be able to make nanogears no more than a nanometer wide. These could be used to construct a matter compiler, which could be fed raw material to arrange atoms and build a macro-scale structure.

However, we are looking at a long period in the future to achieve these nanotechnology goals.

Nevertheless, it is expected to change the way we design and construct everything (e.g., medical equipment, computers, cars, etc.).

Nanotechnology promises to usher in an industrial revolution where machines are the size of molecules. Research in nanotechnology is intensifying. United States government recognized that this is the new frontier and its potential impact is compelling. US national budget allocations for research in nanotechnology have been increasing from $497 million in 2001 to $849 million in 2004, and to several billion dollars.for the next few years.

Nanotechnology also promises to establish an unusual relationship between biology and nanotechnology. There has also been a perception out there right from the beginning that nanotechnology will cause environmental devastation or human disease.

In fact, the US House Committee on Science recently held a hearing to discuss both the Nanotech Research and Development Act of 2003 and the potential societal and ethical implications of nanotechnology. The US Environmental Protection Agency allocated four million dollars in research money for 2003 to the study of environmental impacts of nanotechnology.

Valid concerns about nanotechnology need to be addressed. A team of experts in 2006 warned that the potential of the field of nanotechnology could be compromised unless the scientific community can implement a program of systematic risk research. They have drawn up the list of the following five "grand challenges" in order to evaluate the safety of nanotechnology.

1) Develop instruments to assess exposure to engineered nanomaterials in air and water within the next three to ten years.
2) Create and test ways of evaluating the toxicity of nanomaterials in five to fifteen years.
3) Generate models to predict their possible impact on the environment and human health over the next ten years.
4) Develop ways to assess the health and environmental impact of nanomaterials over their entire lifetime, within the next five years.
5) Organize programs to enable risk-focused research into nanomaterials within the next twelve months.

Nevertheless, we are beginning to reap some of the benefits of nanotechnology today. We will make some predictions about nanotechnology when we discuss the future of technology.

4.5 Frontiers of Computer Technology

After visiting the frontiers of communication and information technology, let us turn to computer technology. Computers are the machines that process information, and it is not surprising that our visit to information science leads us naturally to computers. Today's computers work by essentially manipulating bits that exist in one of two states, zero or one. The power of computation is truly amazing. Three points are worth noticing.

1) It seems that computation can describe all phenomena. We have been able to put every logical argument, scientific equation, and literary work into the basic notation of computation. With the advent of digital signal processing, we can capture video, music, and art in the same digital form.

Recently built programs such as, Kismet by Breazeal, EMIR (Emotional Model for Intelligent Response) by Guerin, and Mehrabian also exhibit primitive feelings. Thus, it appears that computers can even describe emotion.

2) Besides binary states (zero and one) and exact logic, it seems that we can process information using fuzzy logic. Fuzzy set theory, developed by Zadeh in sixties, admits a third probable state. Researchers have now developed fuzzy logic chips for various applications, where the outcome is not exactly definable.

For example, through such chips, we can set threshold values to decide if the clothes are very clean, just clean, or dirty. We might use such logic in our brains while making many decisions. For example, our brain does not solve exact differential equations when we drive and guide our cars through the traffic.

There is an urgent need to reduce the size and increase the speed of computers. Present-day computers mainly use silicon microprocessors. We have been cramming increasing numbers of electronic devices on the microprocessors. Intel Founder, Moore, predicted in 1965 that microprocessors would double in complexity every two years. Following Moore's prediction, the number of electronic devices put on a microprocessor has in fact doubled every eighteen months.

Silicon microprocessors have physical size and speed limitations. If Moore's law is to hold, in just twenty years, we will have to find circuits on a microprocessor measured on an atomic scale. Therefore, we must look for the new computer technology to increase the speed of computations and to reduce the size.

Before we explore the new frontiers, let us review where we stand in terms of the speed and size of computers. The American Blue Gene/L system supercomputer, is one of the world's fastest computers, but records for speed are being broken fast.

The IBM scientists developed their computer at the Lawrence Livermore National Laboratory in Livermore, California. That machine is capable of 136.8 teraflops, or 136.8 trillion calculations per second. Japan wants to develop a supercomputer that can operate at ten petaflops, or ten quadrillion calculations per second, which is seventy-three times faster than the Blue Gene. This is hundreds of thousands times faster than an average desktop computer.

What is the fastest and smallest computer in the world? We carry the fastest and the smallest computer with us. A speed of close to forty trillion operations per second is very fast, but not as fast as our brain, and there is no comparison as far as size is concerned. Our brain is the fastest processor available right now. It is made up of about one trillion cells with one hundred trillion connections between those cells. Roughly speaking, it is handling about ten quadrillion instructions per second.

According to Dr. Oliver Sacks, ("Travels in the Human Mind" – Guardian, October 19, 1996), it is worth remembering that the brain's cortical mantle is the size of a large table napkin when it is spread out. However, the number of possible combinations of nerve connections is hyper-astronomical, of the order of ten followed by several million zeroes.

By contrast, the total number of positively charged particles in the completely known universe is only of the order of ten, followed by eighty zeroes. The potential computing power of our brain is truly amazing!

For the sake of comparison, the present learning software requires twenty minutes of computer training and about one hundred million instructions per second to handle dictation from a single speaker. A good desktop computer is barely able to "understand" speech and take dictation, translating spoken words into written words from one speaker.

Our brain, on the other hand, can understand any number of speakers. It needs no training, and it will make no mistakes. Moreover, speech processing is just one small part of the brain. Our brain also processes complex visual images, controls all functions of our body, understands conceptual problems, and creates new ideas

Following the lead from quantum information processing and from our brain, scientists have been experimenting with both quantum particles and minute strands of DNA to reduce the size and increase the speed of computations.

This gives us two obvious options. We can build quantum computers, which will use particles, atoms, and molecules to perform the computational tasks. Alternatively, we can build DNA-based computers emulating our brain. Both candidates can meet both speed and size requirements. Let us first explore the frontiers of quantum computers.

Quantum computers use particles and nuclei of atoms and qubits instead of bits and use quantum entanglement. Working with atoms as the building block solves the size problem. Working with qubits and quantum entanglement solves the speed problem.

A qubit can contain these multiple states simultaneously, which gives quantum computers their inherent parallelism, and they can work on a million computations at once, unlike our desktop PC that works on one at a time. Quantum entanglement allows us to know the value of a qubit because of its entangled partner without any measurement.

A measurement would destroy the information in superposed states by collapsing to zeroes or ones. A thirty-qubit-quantum computer would have the processing power of a supercomputer that performs ten teraflops (trillions of floating-point operations per second).

There is lot of excitement about such quantum computers. However, as Nelson points out, we need to first understand the general high level principles in quantum information science that govern complex quantum systems such as quantum computers. As he puts it, "These principles relate to the laws of quantum mechanics the way that heuristics for skillful play at chess relates to the game's basic rules."

We are just beginning to understand quantum computing. The technology needed to build practical quantum computer is years away. We see some encouraging signs on the horizon, but to solve real-world problems, we need to build a quantum computer that has at least several dozen qubits. As discussed earlier, nanotechnology will also play an important role in creating a new generation of computer components.

Another approach called spintronics uses the spin of an electron, a quantum property, which is twisted in various orientations to perform logical operations. For example, a magnetic field can change the electron's spin orientation from pointing up to pointing down, by processing the electron like a gyroscope. This corresponds to changing the bit from one to zero. In fact, a spintronic device can create "phits" (phase bits) like qubits (quantum bits) that can have a much wider range of values than just the binary digits, bits zero or one.

The spintronic memory chips would consume very little power, and they would not lose data when power is interrupted since the electron spins stay in their position. In 2009, researchers reported in the *Journal Nature* that they were able to use "spin-polarized" electrons in silicon at room temperature for the first time.

It could lead to computers that require far less power than conventional ones. Moreover, since the effect has been demonstrated in silicon—the material already used in the computer industry—this implies that we could make devices on a commercial scale more easily.

The second choice is DNA computers that can also solve the problem of size and speed. One pound of DNA can store more information than all the electronic computers in the world. More than ten trillion DNA molecules can fit into an area no larger than one cubic centimeter, hold ten terabytes of data, perform ten trillion calculations, and have inherent parallelism. Thus, a small drop of DNA computer, using the DNA logic gates, would have more computational speed than the world's most powerful supercomputer.

We will always have a large and cheap supply of DNA and can make DNA biochips cleanly. Studying DNA computers may also lead us to a better understanding of the human brain. However, a practical and workable DNA computer based on components such as logic gates and biochips is years away. For this to happen, we must understand the workings of our brain.

- "Intelligent" Computer and the Brain

Computers do certain things, like performing complex computations, well. However, computers have difficulty recognizing even simple patterns. This capability is the key to learning and intelligent systems. Even a simple animal brain can perform functions that are currently impossible for computers. Despite developments in computer technology, therefore, our dream of building a true thinking machine remains just a dream.

We simply do not understand the working of human brain. Without the brain, the human body is a chunk of meat. Our brain sets us apart from the animals. How does the brain process information, think, learn, and make "intelligent" decisions? We discuss in detail our brain in the third phase of our journey through life, since the brain gives meaning to our lives.

Our brain can receive and integrate data, extract information, compare it with the stored patterns, and write and modify its own program. It can thus come up with an optimum action plan for different elements of our body, using optimality criteria based on the inherited/acquired values. Our brain is capable of adaptive programming. It can continually modify and update the optimum course of action if the scene and the received information change.

Our brain receives information from senses, associates it with past patterns stored in memory, processes it, uses an optimality criterion according to the values of a particular individual, takes into account the constraints, and comes with the best decision and a plan of action. It also associates feelings and emotions with the received information.

In short, it is the most complex computer. Despite intense research efforts at Microsoft and all around the world, we still do not have a complete understanding of how our brain functions. We need to make substantial progress in these areas.

Building a computer that mimics the brain would involve the creation of connection machines, extensive parallel network processing, and the training of these networks to recognize certain patterns and solve many other specific problems. It would need adaptive loops and self-programming capabilities. It would use words like "behave," "react," "self-organize," "learn," "generalize," and "forget" in place of traditional computing terms. It would perhaps require extensive use of fuzzy logic.

Let us review the limited progress made thus far. We developed, not long ago, the so-called "expert" systems on computers. These expert systems attempted to capture the expertise of a human "expert." Such systems found many industrial applications, especially in areas where expertise was mainly procedure-based and the expert's domain was well defined. An expert system cannot consider options outside its domain.

In other words, it cannot step out of the "box" like human experts and solve a problem when the solution lies outside the domain of computer expert system program. The interest in expert systems, which were supposed to replicate human experts on computer, has been waxing and waning.

After expert systems, the so-called "neural" networks were developed. The artificial neural networks are relatively crude electronic models attempting to emulate the neural structure of the brain. Neurons in the brain can have as many as a thousand inputs, and an artificial neuron can have any number of inputs.

Each of the inputs in the artificial neuron network has a weight that the active inputs contribute to the neuron. Each neuron also has a threshold value. If the sum of all the weights of all active inputs is greater than the threshold, then the neuron is active.

In artificial neural networks, we continuously modify the weights associated with hidden nodes by training the network by giving more and more data, as it becomes available.

These biologically inspired computational networks are thought to be the next major advancement in the computing industry. Despite some progress in "expert" systems and "neural" networks, however, we are still struggling with the clear and technologically meaningful definitions of learning and intelligence.

According to a recent announcement, an "emotionally aware" computer system was featured at a major London exhibition. It is designed to read people's minds by analyzing expressions. The computer, which is connected to a camera, locates and tracks twenty-four facial "feature points," such as the edge of the nose, the eyebrows and the corners of the mouth.

The goal of a recent research project, involving six countries, twenty-five roboticists, developmental psychologists, and neuroscientists, according to its coordinator, Canamero, is to build robots that "learn from humans and respond in a socially and emotionally appropriate manner." The 2.3 million euro project is funded for three years. Despite such projects, we are very far from developing true learning and intelligent machines, comparable to a human brain.

Machines will achieve human-level artificial intelligence by 2029, a leading US inventor has predicted. Ray Kurzweil said in 2008 that humanity is on the brink of advances that will see tiny robots implanted in people's brains to make them more intelligent.

4.6 Frontiers of Control & Robot Technology

Regarding the frontiers of control technology, we have already made tremendous progress in control and guidance of space vehicles, satellites, missiles, and bombs. We have also done wonders in controlling industrial plants and climate inside buildings. However, we have not done well when it comes to controlling the overall environment on our planet Earth. We are still struggling with controlling operational systems, such as economic, social, and political systems.

Finally, despite some progress towards building robots that can perform a few simple tasks, we are very far from building true "learning" and "intelligent" control systems. The remarks concerning thinking machines in the preceding paragraph are equally applicable to intelligent control systems.

For example, to make further significant progress in designing learning and intelligent control systems, we must clearly define the learning process and the intelligence itself. The definition has to be such that we can implement the results using fabricated devices. In other words, before we can design a learning control system we must first understand clearly how a person learns, stores the learned patterns, accesses them, and acts based on these patterns.

There is intense research activity in the area of intelligent control and robotics. For example, the Intelligent Systems Division (ISD) of NIST has an ongoing research and development program, which focuses on addressing both immediate and long-term industry needs.

Such needs include: open-systems architecture standard; intelligent controllers for manufacturing industry and government applications; engineering methodologies and software tools for building intelligent systems; and test methods and metrics for measuring the performance of intelligent control systems.

The leading edge of intelligent systems research is in military robotics, medicine, space, and entertainment, not in manufacturing. To maintain leading edge capabilities in this field, NIST supports the work of other mission agencies. In primarily supports the Army and DARPA in the development of military unmanned ground vehicles, and the Department of Transportation in providing performance measures for crash avoidance systems.

The basic paradigm for intelligent systems control architectures that ISD at NIST is pursuing is hierarchical control. They are pursuing this concept in terms of research to develop a basic understanding of the theoretical issues of intelligent control.

Scientists have expressed concern about the use of autonomous decision-making robots, particularly for military use. These machines could also have negative impacts on areas such as surveillance and elderly care. Autonomous robots are able to make decisions without human intervention.

At a simple level, these can include robot vacuum cleaners that "decide" for themselves when to move from room to room or to head back to a base station to recharge. However, such autonomous machines are also being increasingly used in military applications. Samsung, for example, has developed a robotic sentry to guard the border between North and South Korea.

Robotics research is also intensifying because of the high stakes and space exploration programs. Robots can work in environments hazardous or inaccessible to human beings. Its vast scope and potential applications are illustrated by the following remarks.

Matt Mason, Director of the Robotics Institute at the Carnegie Mellon says, "We take a broad view of what constitutes "robotics." We certainly do build things that you would recognize as robots—robotic museum tour guides, robots for planetary exploration, robots that crawl through pipes and over rough terrain. We build robot arms, mini-factories, grippers, sensors, and controllers. But we also work on speech understanding, and process scheduling, and data mining, and juggling, and traffic safety, and sculpture, and toys, and computer design, and biology, and many more things that don't look exactly like robots."

Robots have been designed for training purposes. For example, the robot "Harvey" at Creighton University is helping student-doctors fine tune their cardiology skills. "Harvey" is a plastic patient that breathes like a human being. Industry is also building robots for public use. For example, Mitsubishi-Heavy has developed "Wakamaru" —a child-shaped humanoid robot that can recognize about ten thousand words. It can work as a house sitter.

The company plans to sell one hundred of these 3.3 feet-tall, sixty-six-pound robots at about $14,300 for residents in central Tokyo. Another Japanese company, Murata Manufacturing, has unveiled a robot named Murata Boy, fifty-centimeter-high weighing five kilograms that can ride a bike. There is bound to be an explosive growth in robot technology in the near future.

4.7 Frontiers of Healthcare Technology

The frontiers of healthcare technology, based on modern physics, are extending at a fast pace. Such technology can monitor our bodies, diagnose problems, treat various diseases, improve the quality of life, and prolong it. In diagnostic tools, we have already moved from X-rays to Computer-Aided-Tomography (CAT Scans), Magnetic Resonance Imaging (MRI), Positron Emission Tomography (PET), and used radiation therapy technology for treatment of cancer.

As regards 3D holograms, the journal Nature reported in early 2008 a breakthrough by US researchers at the University of Arizona, Tucson. Savas Tay and colleagues developed a novel material in which one can create holographs in minutes. The images that the material captures are almost as sharp as those broadcast on US television.

We can construct large screens using this polymer, which opens up many possible uses for the 3D images. The material can remain stable throughout hundreds of write and erase cycles. The ability to refresh images in holographs could mean that surgeons could use them as a guide during operations, or they could be used as a better way for pharmaceutical researchers to study molecular interactions for new drugs during simulations.

Two types of cutting-edge technology are particularly promising. One is the biotechnological revolution, where we can manipulate life at the cellular level. The other is nanotechnology, where we have the ability to manipulate matter precisely on the atomic level. The merger of these two technologies holds great promise.

Having discussed the frontier of nanotechnology, we now explore and focus on the frontier of biotechnology. We have already mapped the human genome and we are at the threshold of remarkable discoveries that will undoubtedly prove to be extremely important for developing crucial healthcare technology.

The genetic revolution has just begun. We should soon reap the fruits of research in this area. We shall discuss the progress in the field of genetic and its impact on health technology, when we visit the field of genetics during the third phase of our journey through life.

Such studies and efforts in the field of healthcare science and technology are bound to lead to treatment of various diseases. Stem cell research, as discussed in the third phase of our journey, holds great promise. Stem cells are the subject of intensive research for obvious reasons. There is controversy about using stem cells from discarded embryos. However, scientists are finding ways to harvest stem cells by other methods.

For example, biologists have developed a technique for establishing colonies of human embryonic stem cells without destroying embryos. If confirmed in other laboratories, it would seem to remove the principal objection to stem cell research. Scientists have also succeeded in making ordinary skin cells of a mouse behave like embryonic stem cells by manipulating four genes. However, similar results for human cells might be years away.

The stem cells have the potential of regenerating failing body parts and curing diseases not responding to drug treatment. For example, sections of liver were recently created using stem cells from umbilical cords by a team at Newcastle University. British scientists announced in 2007 that they have grown part of a human heart from stem cells for the first time. The heart surgeon, Yacoub, who led the team, said doctors could be using artificially grown heart components in transplants within three years.

Chapter 5
Future of Technology

5.1 Introduction

Predicting the future is always a risky business. Predictions are difficult to make and they are usually off the mark. In this chapter, we visit some interesting earlier predictions in technology that failed to materialize. Then, we go out on a limb and make some predictions about the future of technology.

We predict some remarkable developments in technologies, such as nanotechnology, quantum communication, quantum and DNA-based intelligent computers, smart robots and control systems, teleportation—the stuff in *Star Trek*, and the new diagnostic and treatment technology for healthcare. In fact, the future developments would make the present technology look primitive. We end our journey with some final thoughts.

5.2 The Impact of Technology on Environment

We all agree that rapid industrialization and spread of technology has contributed to releasing pollutants into the atmosphere. In other words, technology has had an impact on the global environment. How endangered is our planet due to our actions in polluting the environment? Although the spread of technology and rapid industrialization has had an impact on climate, all scientists and climatologists do not agree on the degree of damage we have caused.

One area of considerable concern and debate is the impact of greenhouse gases, such as carbon dioxide, methane, and nitrous oxide released into the atmosphere by us. Simply stated, the greenhouse gas is relatively transparent to light from the sun, but it can absorb some thermal radiation from the earth. Normally, the earth balances the incoming solar radiation by emitting the thermal radiation. However, the presence of greenhouse substances can absorb some thermal radiation, inhibit cooling, and lead to some warming.

Should we be alarmed by the prospect of global warming? Some recent studies suggest that we must mend our ways. In 2009, Dr. Foley reported in the journal Nature the results of a team study with twenty-seven experts from around the world. The team decided to look at the Earth's safe operating space, and study the potential risks that could push our planet to a state of instability.

The researchers defined nine categories of risk within that space, including global warming, ocean acidification, and stratospheric ozone levels. The scientists tried to quantify the risks by compiling and analyzing published work in their particular areas of expertise. Their study suggests that we have already pushed the planet too hard in at least three ways.

According to Dr. Foley, the changes in the environment on a global scale would change this planet into something we have never seen in all of human history. It is not the end of the world, but it is the end of the world, as we know it. Foley said, "We think that a little damage to the environment is OK, but at some point, the planet just cannot take it anymore, which is especially true when it is taking multiple hits at once."

Their study suggests that we have already pushed the planet too hard in at least the following three ways:

1) The current CO_2 concentrations measure about 387 parts per million, but in this study the climate researchers estimate that nature will remain in balance only as long as carbon dioxide concentrations in the atmosphere remain below 350 parts per million.

2) Currently, species are already disappearing at a rate of 100 species per million, and the projected rates are ten times higher than the present rate. However, biodiversity researchers estimate that species loss is sustainable only if we lose fewer than ten species for every million on Earth.

3) We are also dangerously close to the thresholds of safety for freshwater use, ocean acidification, and the conversion of forests and other ecosystems into farms and cities. Crossing those lines may lead to a cycle of global catastrophic change. Nitrogen outputs from chemical fertilizers and other human activities are already threatening to damage irreparably freshwater and marine ecosystems.

The specific numbers in the study remain estimates, and it should start a debate, according to Steve Carpenter of the University of Wisconsin, Madison. This study points out, nevertheless, an important issue. We must recognize that we cannot just keep abusing the planet forever.

Otherwise, we might reach a point of no return, and destroy the delicate balance in nature forever. Of course, nature and our planet would find another state of equilibrium and balance, but it might not support human life.

According to another recent study, scientists warn that more than a third of species, assessed in a major international biodiversity study, are threatened with extinction 17,291 were deemed to be at serious risk of extinction out of the 47,677 species in the IUCN Red List of Threatened Species. These species included 21% of all known mammals, 30% of amphibians, 70% of plants, and 35% of invertebrates.

However, according to some scientists, these large-scale computer simulations may be overestimating the impact of climate change on biodiversity in some regions. They think that the models that analyze vast areas do not take into account local variations, such as topography and microclimates.

It is noteworthy that several prominent scientists do not agree with the hypothesized impact of greenhouse gases, such as, carbon dioxide, methane, and nitrous oxide released into the atmosphere by us. According to Dr Lindzen, Professor of meteorology at M.I.T, even a doubling of CO_2 would only upset the original balance between incoming and outgoing radiation by about 2 percent. He strongly disagrees with the conclusions drawn by the U.N.'s Intergovernmental Panel on Climate Change (IPCC) on the adverse impact of greenhouse gases.

According to him, the measurement used in the thousand-page IPCC Report, the globally averaged temperature anomaly (GATA), is always changing. Sometimes it goes up, sometimes down, and occasionally—such as for the last dozen years or so—it does little that can be discerned.

According to a statement issued after the last IPCC Scientific Assessment two years ago, it is likely that most of the warming since 1957 (a point of anomalous cold) was due to man. This claim was based on the following weak argument. The current models used by the IPCC could not reproduce the warming from about 1978 to 1998 without some forcing, and the conclusion drawn was that the only possible force at play was man.

The argument assumed that these models adequately deal with natural internal variability—that is, such naturally occurring cycles as El Nino, the Pacific Decadal Oscillation, the Atlantic Multi-decadal Oscillation, etc.

However, the major modeling centers acknowledged that the failure of these models to anticipate the absence of warming for the past dozen years was due to the failure of these models to account for this natural internal variability. Thus, even the basis for the weak IPCC argument for anthropogenic climate change (changes due to CO_2 emission) was shown to be questionable.

According to Dr Lindzen, the notion that complex climate "catastrophes" are simply a matter of the response of a single number, GATA, to a single forcing, CO_2 (or solar forcing for that matter), represents a gigantic step backward in the science of climate.

Many disasters are claimed to be evidence of warming, when these disasters are simply normal occurrences. Their occurrence involves complex phenomena, which are dependent on the confluence of many factors.

According to Dr Lindzen, the normal occasional occurrences of open water in summer over the North Pole, droughts, floods, hurricanes, sea-level variations, etc. are all taken as omens, portending doom due to our sinful ways (as epitomized by our carbon footprint). All of these phenomena also depend on the confluence of multiple factors, and not just CO_2 increases.

It makes sense that one single factor, such as greenhouse gases, is not entirely responsible for climate changes. Furthermore, we do not completely understand the complex interplay of various factors and their effect on the climate changes. Thus, our computer models are far from perfect. Nevertheless, it is also true that the climate changes do affect our planet and species. It is also true that we are contributing to the climate changes to some extent because of rapid industrialization.

At present, we might not be able to quantify the effect of our contribution to polluting the environment. Not all of us might agree on when we would reach the limits where greenhouse gases or other pollution would seriously affect the planet and our way of life.

However, we must all agree that we can reduce pollution through our efforts. It is wise to err on the side of safety. It would thus be wise to take all the steps we can to minimize our impact on the environment. We can certainly do more in this area.

We can improve the existing structures and industrial plants, and build new structures and industrial plants that would conserve energy, and release less greenhouse gases and other pollutants into the atmosphere. In these efforts, we can make technology our friend, and develop processes that minimize the impact of industrialization on the environment.

Let us take an optimistic view, and hope that the world community will take steps to reverse or at least prevent further deterioration of conditions on our planet. If we would not push earth into a state of instability, new scientific and technological discoveries promise a better future for technology and for humanity.

5.3 Past Predictions for Technology

Knowing the past trends, we could perhaps project what would happen in the near future. However, one can never foresee the sudden significant abrupt changes in science and technology. Sudden future outburst of unforeseen discoveries in science could quickly change the direction and development of technology. A quick look at the following wrong predictions confirms this fact:

1. No flying machine will ever fly from New York to Paris ... [because] no known motor can run at the requisite speed for four days without stopping... -- Orville Wright

2. "Heavier-than-air flying machines are impossible." — Lord Kelvin, British mathematician and physicist, president of the British Royal Society, 1895.

3. "A rocket will never be able to leave the Earth's atmosphere." — *New York Times*, 1936.

4. "There is little doubt that the most significant event affecting energy is the advent of nuclear power ... a few decades hence, energy may be free -- just like the unmetered air...." -- John von Neumann, scientist and member of the Atomic Energy Commission, 1955.

5. "Space travel is utter bilge." -- Dr. Richard van der Riet Woolley, Astronomer Royal, and space adviser to the British government, 1956. The next year Sputnik orbited the Earth.

6. "There is not the slightest indication that energy will ever be obtainable from the atom."—Einstein

7. "When the Paris Exhibition [of 1878] closes, electric light will close with it and no more will be heard of it." —Oxford professor, Erasmus Wilson

8. "Fooling around with alternating current is just a waste of time. Nobody will use it, ever." — Thomas Edison, American inventor, 1889 (Edison often ridiculed the arguments of competitor George Westinghouse for AC power).

9. "There is no reason anyone would want a computer in their home." — Ken Olson, president, chairperson, and founder of Digital Equipment Corp., maker of big business mainframe computers, arguing against the PC in 1977.

10. "X-rays will prove to be a hoax." — Lord Kelvin, President of the Royal Society, 1883.

Predicting the future of technology can indeed be hazardous (Fig. 5.1).

Predicting Future of Technology

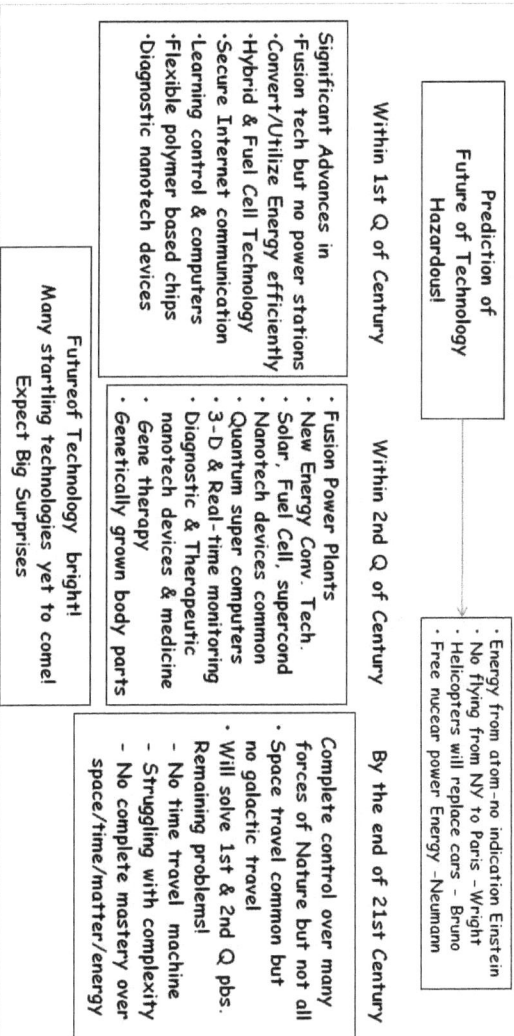

Prediction of
Future of Technology
Hazardous!

- Energy from atom–no indication Einstein
- No Flying from NY to Paris – Wright
- Helicopters will replace cars – Bruno
- Free nucear power Energy –Neumann

Within 1st Q of Century

Significant Advances in
- Fusion tech but no power stations
- Convert/Utilize Energy efficiently
- Hybrid & Fuel Cell Technology
- Secure Internet communication
- Learning control & computers
- Flexible polymer based chips
- Diagnostic nanotech devices

Within 2nd Q of Century

- Fusion Power Plants
- New Energy Conv. Tech.
- Solar, Fuel Cell, supercond
- Nanotech devices common
- Quantum super computers
- 3-D & Real-time monitoring
- Diagnostic & Therapeutic nanotech devices & medicine
- Gene therapy
- Genetically grown body parts

Future of Technology bright!
Many startling technologies yet to come!
Expect Big Surprises

By the end of 21st Century

Complete control over many forces of Nature but not all
- Space travel common but no galactic travel
- Will solve 1st & 2nd Q pbs. Remaining problems!
  - No time travel machine
  - Struggling with complexity
  - No complete mastery over space/time/matter/energy

Fig. 5.1 – Predicting the Future of Technology

5.4 Predictions for the Next Quarter-Century

Sandberg said that the technological progress could not be predicted. On the other hand, in the June 1994 issue of *IEEE Spectrum,* Miller said that, "predicting the distant future is easy. Solar power will be harnessed, cures will be found for present-day diseases, and the New York Mets will win the World Series. The real difficulty is predicting tomorrow - figuring out the next step in getting from where we are to where we are going."

Despite these interesting observations on predictions, people still keep predicting the future. Carrying on this tradition, we indulge in this exercise and predict, or more appropriately, speculate on, the future trends in technology.

One of the major revolution in technology would be through the merger of various technologies. Such merger of the communication, computers, and control technologies will produce astonishing new gadgets.

The day is not very far when we will have a book size device, which has a 3-D screen, video display, telephone and televideo conferencing, projection, access to all TV programs, etc.. We already have the chips and other technology, and iPAD is an example of such devices. The advances in this area would only be limited by human imagination.

In my opinion, during the first quarter of this century, the following developments are likely to happen.

1) Our daily life would become even more dependent on technology, as technology plays an increasingly dominant role in every occupation.

2) We will still not have new technology available for the safe disposal of the nuclear waste from nuclear fission.

3) We will make significant advances in fusion technology, but it will not be ready for safe power generation.

4) We will continue to use the existing energy conversion technologies more efficiently, and develop new energy conversion techniques. Solar power, wind power, and other alternative renewable energy resources will be developed and used increasingly.

5) Hybrid techniques, such as gas and electric power, would become common for cars. We are already seeing such cars on the road. Hydrogen fuel cell technology will mature, and hydrogen-fuel driven vehicles and hydrogen fuel stations will be available.

6) We will develop new superconductor materials, operating at relatively higher temperatures (but well below room temperature), which will find many applications.

7) Nanotechnology will advance with applications in consumer electronics through miniaturization,

in chemicals and basic materials through new materials, and in energy through thin film photoelectric and fuel cells.

8) In materials, we will develop technology to build new materials with almost any properties that we desire.

9) In communication, we will develop new means to transmit securely more data at a faster rate. In modern homes, all the communication systems, including entertainment systems, will be interconnected through computers. Mobile phones would connect people in unimaginable ways.

10) In control, we would develop the learning control systems—but not "intelligent" control systems, like human beings. The application of GPS technology in missile guidance and navigation is already producing amazing results.

11) In computers, over 90 percent of the world population will own some type of computer. Computer applications will increase dramatically, and a PC with advanced operating systems will link and control all media communication and other home accessories. We will be able to access easily all the human knowledge on the Internet, as search engines become smarter and more powerful.

12) The size of computers will continue to shrink and the computational speed will continue to improve. Semiconductors and silicon chips

would be on the way out, replaced by polymer based flexible semiconductor devices and chips. A major drive for research in nanotechnology will be undertaken to find new ways of making microchips for computers. Despite significant advances in quantum information science, we will not have the quantum, DNA computers, or thinking machines on the market.

13) In healthcare, we will develop several new devices and tests to monitor and diagnose various diseases. Nanotechnology will provide new tools for sensing, diagnostic, and new therapeutic nanomedicines. We will develop technology to take a clear live dynamic video of the inside of our body.

14) We will make great strides towards mapping the functions of different genes in the human body, and will be farther along the road to developing gene therapy. WE will have many gene-based medicines to treat various diseases. Growing various human body parts from cells and their use will become possible.

15) Existing technologies for global ATM networks, online banking, debit cards, and paperless checking applied to healthcare would produce astonishing results. Online physician appointments, online access to personal updated medical records, home-based diagnostic devices communicating with doctors, online

prescriptions, and instant access to specialists will become widely available.

5.3 Predictions for the Next Half-Century

By the *middle of the twenty-first century*, we are likely to achieve the following.

1) We will have commercial power plants based on fusion technology, and fission power plants will be outmoded.

2) We will have practical transducers to convert any form of energy to another, and new energy conversion techniques will replace the existing technology.

3) Use of solar energy, windmills, hydrogen fuel cells, and superconductors will be wide spread.

4) We will have personal transport vehicles traveling in the sky.

5) Nanotechnology will be mature, and we would be using many devices and new materials based on this technology. However, we still will not have the technology to make tiny robot and will not face the "gray goo" problem—a world overrun by tiny robots.

6) Quantum information science will be fully developed, and quantum communication will be common. Teleportation will still be in the research stage. Quantum computers, using qubits instead of bits, will replace the present computers based on silicon chips.

7) People will carry supercomputers in their pockets that could access any information or contact any individual from any place.
8) DNA computers and thinking machines will still be in the research and development stages.
9) In control technology, use of intelligent robots for mundane tasks will be quite common. Applications of control systems that learn from past patterns in memory will also be widespread.
10) Research on merging biotechnology with nanotechnology will enter an advanced stage with several applications in healthcare. Delivery of medicine to specific targeted cells using nanotechnology will become common.
11) In healthcare technology, monitoring and diagnosing various diseases with exact 3-D pictures of the inside functioning of the body will become routine tasks. Diagnosing and treating the cause of a disease will replace the present system of diagnosing and treating symptoms with drugs.
12) Research on human genome and gene mapping will have borne fruits, and gene therapy for various diseases, including cancer, will become widespread. Using gene therapy, scientists have already successfully treated patients with hemophilia and infants with the rare "bubbly boy" disease, who have no immune system.
13) Growing and using human body parts will also be quite common, and it will be possible to have

offspring with any traits we desire through manipulations of the sperms and eggs.

14) Unfortunately, human behavior will not change much. We will continue to develop more powerful and "smart" destructive weapons with amazing accuracy. Our planet will still be a dangerous place to live on, as new technology will provide more powerful Weapons of Mass Destruction (WMD) or arms and ammunition for Mutually Assured Destruction (MAD).

5.4 Predictions Beyond

The crystal ball for predicting the future of technology gets hazy beyond this point. In one sense, it is easy to have long-range predictions. We can say that all the preceding predictions will most likely have come true by the end of the century. However, the future course and the exact developments in technology are more difficult to predict during the second half of the twenty-first century.

Nevertheless, we go out on a limb and boldly make some predictions about what we might or might not be able to achieve by the end of this century.

1) By the end of the twenty-first century, we will have better answer to the question: How can I be more comfortable? We will control our environment to an extent that is hard to imagine today. We will have gained gain mastery over many forces of nature. We might even find some

means to divert hurricanes and other destructive forces of nature to less populated area.

2) By the end of this century or in the near future, however, we will not gain complete mastery over space, time, matter, and energy.

3) We will not be able to travel to other stars and galaxies because of the light speed limitation. We will not be able to stop the flow of time or have time-travel machines.

4) We will not be able to create matter from energy on a large scale, and we will still be harnessing an extremely minute amount of energy in our universe.

5) We will still be struggling with the problem of complexity, which involves understanding how nature builds extremely complex systems.

These predictions may be optimistic or pessimistic depending on one's outlook. Who knows? For example, who could have predicted twenty-five years ago the present revolutionary developments in communication and computer technology?

Judging from experience, developments in science and technology come in sudden spurts, and surprise us. Many technologies came and disappeared quickly. Then there were those technological developments that nobody expected, that stuck with us and changed our lives.

There will be many unexpected turns and developments in technology that will open new venues for research and development. At the end of twenty-first century, we will still be marching on toward new frontiers of technology, provided we can avoid building and using more arms for mutually assured destruction (MAD) and other undesirable byproducts of newly emerging technology.

5.5 End of the Journey

We come to the end of our journey through technology. There is no question that modern technologies have brought an incredible number of benefits to human societies. However, they have had—and are having—a major influence society and the environment.

Examples of impacts on the environment include not only global warming and large-scale deforestation, but also the effects of industrial-scale fishing, sea bottom trawling, etc. To some extent, it has also contributed to an increasing disparity between the haves and have-nots—through a kind of 'pull apart' effect.

Immediate benefits from the application of technology are so great in many so-called industrially developed countries—at least amongst the ruling classes—that a sense of arrogance seems to prevail.

They believe that technology and big government can solve all possible problems of society. This confidence might prove to be misplaced. Obviously, there are limits to what big government and technology can do to solve today's problems.

Our existence on this planet, and especially the time we have spent on developing technology, does not even show up as a blip on the time scale for the existence of the universe. We have yet to discover and conquer many new frontiers in science, and develop many new technologies. Many of us may not be around to witness as most of the limitations of science and technology are overcome in years to come.

Let us hope that the forthcoming advances in science and technology will make us better human beings. During our journey onwards, we should learn to preserve the earth's environment, and not destroy our planet though pollution or weapons of mass destruction. Protecting our planet is very important for humanity.

Our planet earth, like any system, has a certain operating range of parameters for keeping it in a stable condition for the human species to survive. If we change adversely the environment or ecological balance, and push the planet into a state of instability, the human race might not survive. Recent studies do suggest that we must mend our ways. We must recognize that nature has hard edges, and we cannot just keep abusing the planet forever.

We could end our journey through technology on this somber note. However, let us take an optimistic view and assume (A big assumption, since Einstein said that human stupidity is infinite!) that the recent efforts by the sane world community to reverse, or at least prevent further, deterioration of conditions on our planet will succeed.

If we could develop new technology for keeping earth within the safe operating window and not push it into a state of instability, new scientific and technological discoveries promise a better future for us.

Let us sincerely hope that human intelligence will keep evolving, and that a bright future lies ahead of us. We shall explore the minutest segments and the farthest edges of our universe, and use new scientific discoveries to develop more technology.

Finally, I hope the readers enjoy this book as much as I enjoyed writing it.

***********End of the Journey**********

For any feedback, please connect with me on Facebook or my website:
www.knowingtheunknownbooks.com

INDEX

U

www.ingramcontent.com/pod-product-compliance
Lightning Source LLC
Chambersburg PA
CBHW060032210326
41520CB00009B/1092